Data Analytics for Corporate Debt Markets

Data Analytics for Corporate Debt Markets

Using Data for Investing, Trading, Capital Markets, and Portfolio Management

Robert S. Kricheff

Vice President, Publisher: Tim Moore
Associate Publisher and Director of Marketing: Amy Neidlinger
Executive Editor: Jeanne Glasser Levine
Development Editor: Natasha Torres
Operations Specialist: Jodi Kemper
Cover Designer: Alan Clements
Managing Editor: Kristy Hart
Senior Project Editor: Betsy Gratner
Copy Editor: Karen Annett
Proofreader: Williams Woods Publishing
Indexer: Tim Wright
Compositor: Nonie Ratcliff
Manufacturing Buyer: Dan Uhrig

Published by Pearson Education, Inc.
Upper Saddle River, New Jersey 07458

For information about buying this title in bulk quantities, or for special sales opportunities (which may include electronic versions; custom cover designs; and content particular to your business, training goals, marketing focus, or branding interests), please contact our corporate sales department at corpsales@pearsoned.com or (800) 382-3419.

For government sales inquiries, please contact governmentsales@pearsoned.com.

For questions about sales outside the U.S., please contact international@pearsoned.com.

Printed in the United States of America

First Printing February 2014

ISBN-10: 0-13-355365-5
ISBN-13: 978-0-13-355365-9

Pearson Education LTD.
Pearson Education Australia PTY, Limited.
Pearson Education Singapore, Pte. Ltd.
Pearson Education Asia, Ltd.
Pearson Education Canada, Ltd.
Pearson Educación de Mexico, S.A. de C.V.
Pearson Education—Japan
Pearson Education Malaysia, Pte. Ltd.

Library of Congress Control Number: 2013953301

I would like to dedicate this book to my wonderful mother-in-law Hope Mullen Bowe and also to the memories of my father-in-law Dr. Edward T. Bowe and my Uncle Bill Glou, both of whom were great men with wonderful souls. All of them have made my life, and many others, so much better.

Contents

About the Author

Robert S. Kricheff (Bob) is a senior vice president and portfolio manager at Shenkman Capital Management. Before joining Shenkman Capital, he worked for more than 25 years at Credit Suisse in Leveraged Finance. Prior to leaving Credit Suisse, he was a managing director and head of the Americas High Yield Sector Strategy.

He has worked doing credit analysis in several industries, including media, cable, satellite, telecommunications, health care, gaming, and energy, and has worked with corporate bonds, loans, convertibles, preferred stocks, and credit default swaps as well as emerging market corporate bonds. He has also run strategy and has overseen portfolio analytics.

Bob is the author of *A Pragmatist's Guide to Leveraged Finance: Credit Analysis for Bonds and Bank Debt* and two e-book shorts, *The Role of Credit Default Swaps in Leveraged Finance Analysis* with Joel S. Kent and *How to Analyze and Use Leveraged Finance Bonds for Project Finance*, all published by FT Press. He also contributed to the book *High-Yield Bonds: Market Structure, Valuation, and Portfolio Strategies* by Theodore M. Barnhill Jr., William F. Maxwell, and Mark R. Shenkman, published by McGraw-Hill.

Bob graduated from New York University School of Arts & Science with a BA in journalism and economics and received an MSc in financial economics from the University of London School of Oriental and African Studies.

About the Contributors

Jonathan Blau is a managing director of Credit Suisse and head of Global Leveraged Finance Strategy within the Investment Banking Division, based in New York. In this role, Mr. Blau is the strategist for leveraged finance, covering the drivers of risk and return for high-yield and leveraged loans in the United States and Europe.

Jonathan joined Credit Suisse in November 2000 when the bank merged with Donaldson, Lufkin & Jenrette (DLJ), where he was a senior vice president in High Yield Research. Jonathan joined DLJ in 1997 from Chase Securities. Prior to Chase, he worked at First Boston, which he joined in 1992.

Before he began his career in finance, Jonathan was a computer design engineer at Data General, Tandem Computers, and Alliant Computer Systems. He holds five U.S. patents, and his work at Data General was featured in the book *The Soul of a New Machine* by Tracy Kidder.

Mr. Blau received a BS in computer science and engineering from the Massachusetts Institute of Technology.

Miranda Chen is a member of the US Credit Strategy team at Credit Suisse based in New York. Her work covers both investment-grade and high-yield markets and spans cash, credit derivative, and structured credit trading strategy, incorporating fundamental, structural, and technical analysis as well as macro and cross asset themes. Previously, Miranda was a member of the Leveraged Finance Strategy team at Credit Suisse where she covered high-yield, leveraged loans, and collateralized loan obligations (CLOs). Miranda is a Chartered Financial Analyst (CFA) Charterholder and holds a dual degree from the University of Pennsylvania in economics (finance) and electrical engineering.

Alexander Chan joined Shenkman Capital in 2010 as vice president of Quantitative Strategy and Portfolio Analytics. Prior to joining Shenkman Capital, Mr. Chan was the leveraged finance strategist at Nomura Securities publishing on high-yield bonds, leveraged loans, and CLOs. He also had a similar role at Barclays Capital for several years prior to Nomura. Mr. Chan spent eight years at Credit Suisse/Donaldson, Lufkin & Jenrette (DLJ), where he was a member of the Institutional Investor ranked Leveraged Finance Strategy team. During his time at Credit Suisse, Mr. Chan held a variety of roles, including publishing strategist, advising institutional investors on manager selection, and CLO sales. Mr. Chan received a BA from Tufts University in quantitative economics and international relations.

Section I
Introduction to Data Analytics for Corporate Debt Markets

By Robert S. Kricheff

This book introduces and outlines the main tools and methods that are currently used in data analytics in the corporate debt market. It discusses some of the different end users of these analytics and how they typically use (or don't use) these tools in the corporate debt markets today.

Although fundamental credit work is still critical to proper investment selection, data analytics helps improve and speed up the process of credit selection and helps to guide market participants as to where they should focus their credit work, the same way a radiologist might say that an X-ray helps him to draw an *X* to show a surgeon where to cut.

At its best, these analytics can be a forward-looking tool guiding you on how to be positioned in the future. It also is a strong tool to use in analyzing how a portfolio or trading desk might be currently positioned relative to the state of the marketplace or what part of its holdings impacted past performance. Markets shift rapidly, and analytics helps to monitor those shifts, both on a relativve value basis and within the makeup of a given market.

This book takes you through some of the terms and tools used in the market, including indexes and databases, and then outlines how analytics is used to compare markets and develop investment themes and then to pick out debt issues that fit or do not fit those themes. It also covers how investors try to analyze short-term supply and demand and covers some special parts of the market that utilize analytics.

The book is definitely *not* a guide on how to build programs or systems, but should be a guide in some of the ways analytics is being used in the debt markets and lend some insight into how it may be used in the future.

1

The Basics

Why Use Analytics?

Using data analytics is absolutely necessary in the modern corporate debt markets; the sheer growth and complexity of the market make it almost impossible to do any major role in these markets without at least some use of analytics. It is required to compete in the modern markets and, with increasing focus on managing risk, it is a necessary tool to manage large trading desks at investment banks and portfolios at money management firms. The corporate debt markets are dynamic, not static, and analytics is necessary to see how the markets are changing.

New entrants into this market should be aware of the basic tools used for analyzing data in the markets, and more seasoned practitioners should also be aware of how others in the markets may be using analytics. Everyone should be looking at new ways to analyze the markets and be looking for ways to use new developments in the market in his or her analytics.

From risk managers, to credit analysts, portfolio managers, investment bankers, capital markets and syndicate personnel as well as traders, salespeople, and asset allocators within the multitrillion dollar international corporate debt market, if you do not use data analytics in your business and investing decisions, you will be at a massive disadvantage.

Typically, numerous programmers, system designers, and managers work on building and maintaining these systems. If they want to excel, they also should have a strong understanding of the unique nature of corporate debt products and the type of analysis that end users want to undertake.

What Is Data Analytics?

What is data analytics? It consists of gathering the proper data and then manipulating and examining it so that you can reach logical conclusions about the data you are looking at. In the case of this book's topic, it involves data about the corporate debt markets and seeing what conclusions you can derive about the various markets relative to each other and the subsets of each market and their relationships. Ideally, you will see trends developing over time or relationships that do not seem to make sense and these can create opportunities. Several products have been developed in the marketplace, which depend on these analytics as well.

Many tools are used in analytics and you can try to achieve many different goals with the analytics. This book is an introduction that focuses on data and analytics used in the corporate debt markets. It is not intended to be a detailed how-to book, but it does outline how analytics is typically used by the major players in the corporate debt markets as well as the major tools and methods currently used in data analytics in these markets. This book sheds some light on ways these tools are designed, points out some of their shortcomings, and offers a glimpse into where analytics in this field may be starting to go in the future.

Many of the analytical tools used are similar or even exactly the same as those used in the traditional government debt market. Also utilized are many features from the equity markets. Blending these features and detailed macro- and microanalysis creates unique analytical tools for corporate debt instruments.

How Data Analytics Is Used and How It Differs from Credit Analysis

If the fundamental credit analysis, which is so important in making proper investment decisions, starts from the bottom up, data analytics can best be viewed as working from the top down. The work can start from examining some of the macro data on the various fixed-income markets and comparing data across these markets. The next step down the ladder would be to start examining the various subsets of a given market. All of this analysis should encompass return analysis as well as comparisons of relative value and volatility. You will likely run this analysis through historical cycles and be trying to extract how past trends might give you a road map into the future or what relationships seem out of line from historic trends and present opportunities or risks. You may spend a particular amount of time examining specific periods of heightened volatility and how various market segments reacted during these times.

This macro work will help you to develop themes and strategies that you and your team will need to follow in your investing or trading strategies. It will also help you understand where your current strategies have you positioned. This may all lead to your developing investment themes you want to pursue.

To find the specific investment ideas that you want to delve into, you will likely use analytics on a database using queries and data sorts to develop narrower and narrower lists of securities that meet the criteria for your investment themes. One of the keys to the quality of the analytics is how robust the database is.

After you develop these lists, the detailed fundamental and relative value analysis can kick in for credit selection. A portion of the analysis that you need to do to best understand when and where to invest includes technical analysis of supply and demand in the marketplace. Many of the items that you will use for your analysis have their own idiosyncrasies that you need to understand to best do your analysis.

One of the most widely used tools is market indexes, which each have benefits and drawbacks that need to be understood. It is also critical to understand credit default swap (CDS) indexes, exchange-traded funds (ETFs), and collateralized loan obligations (CLOs), all of which rely heavily on data analytics to be run and managed. In doing all this analysis, you need to understand who the end user is as well as the other players who can influence the marketplace. You also need to understand the basics of bond math as well as statistics—both of which are necessary for undertaking this work.

If much of the analytic work just described is to reach a conclusion about where you should be best positioned for optimal performance, performance attribution shows you how you have been positioned and how that has impacted you. Performance attribution analysis is a culmination of much of the data analytics tools that are used across the marketplace.

An Example

This example starts with a look from the top down, starting with some macro analytics, and then brings it down to a more micro level of actual credit selection possibilities that can fit the theme developed by the macro analytics.

A macroanalysis might include looking at how the average yields on the leveraged loan market are trading relative to the average yields in the high-yield bond market versus historic trading patterns. In this analysis, you would consider many factors, including the following:

- The bulk of the loan market has floating-rate coupons while the high-yield bond market does not.
- During various historic cycles, what was the economy doing versus the current environment?

- During various historic cycles, what direction were interest rates going versus the current environment?
- How has the makeup of the two markets changed over time? (For example, does the high-yield market now have significantly more secured bonds than it did during some other cycle?)

In addition, you would likely consider and address many other factors in the analysis.

In the preceding example, let's assume you see a historical pattern that during a period of soft gross domestic product (GDP) growth, you actually see that yields on the high-yield bond market go up materially more than on loans. You then might want to take this to another level and see if this pattern is true for all bonds or just certain types. Perhaps it is most pronounced in notes that have a subordinated ranking and that were issued by companies that are in cyclical industries.

If you are worried about entering a period of soft GDP, you might then want to analyze which bonds fit the criteria to make sure you do not have exposure to them. You can then use a database with a query system to develop a list of debt issuers that are in cyclical industries and have subordinated bonds outstanding. You might then want to take this list of subordinated bonds, compare it with the spread on the same company's loans, and compare this spread to historical trends while also having a credit analyst explore the overall credit strength of the company.

This is just a simple example of how data analytics can be used in the corporate debt market.

I believe that analytics in the corporate debt market is still in the early stages of development and usage. There will be significant advances in the use of analytics going forward and this will lead to an increase in spending in coming years. Money will need to be spent to improve risk management and develop new analytical tools for the markets. Data management and analysis techniques that are being used in other fields, such as in marketing, are likely to be adapted for these markets.

How This Book Is Structured

Section I outlines why the corporate debt markets are so different from other asset classes. This section highlights some of the problems and difficulties with trying to undertake analytics in the corporate debt markets. It also discusses some of the common data sources.

Section II goes over some of the key terms used in the marketplace and common to analytics and also reviews typical tools that are used in the markets. If you are familiar with the markets, you might choose to skim this section. If you are relatively new to the markets or are coming from the programming and system side, you might not be familiar with the topics covered in this section and should find it helpful.

Section III summarizes the definitions of the various markets within the overall spectrum of corporate debt. It also outlines who the major players are in the corporate debt market and how they utilize data analytics.

Section IV covers indexes. Jonathan Blau discusses the details and design of corporate debt market indexes. Indexes are the most widely used source for performance comparisons of various markets and for portfolio performance attribution and comparison. Understanding how these indexes work, the different methodologies, and their shortcomings is key in understanding much of the everyday analytics that goes on regarding market data.

Section V examines how data analytics is used starting from a top-down approach. This macro approach starts with examining performance and relative value at the market level and then works its way down to analyzing key subsectors of the markets to develop investment themes and capture trends that might be occurring within a given market. We then explore some of the tools used to derive lists of possible credits to select that can meet the investment themes that are developed.

Section VI focuses on analyzing supply-and-demand trends in the marketplace, known commonly as technicals. Understanding the trends in market technicals can be critical in helping to make timing decisions and weighting decisions in the marketplace.

Section VII explores special vehicles that have evolved in the market. Miranda Chen, an expert on these products, authored this section, which outlines liquid bond indexes, credit default swap indexes, and exchange-traded funds and shows how they depend on analytics for their construction. This section also outlines why monitoring these vehicles can help give insight into market trends and technical more quickly than some other sources of data.

Section VIII explores collateralized loan obligations. These structured products also utilize analytics systems to be structured and to maintain their portfolios within the rules that they have to operate. Similar to the other structures' vehicles, understanding and monitoring data on these products can add insights into analyzing trends in the corporate bank loan markets.

Section IX outlines the key features of portfolio analysis and performance attribution. This is one of the most developed uses of analytics in the market and we would expect to see the use of such products expand and evolve. This section is written by Alexander Chan, who has worked on developing and running several of the early attribution systems and some of the most current and up-to-date systems.

Section X takes a look at some of the possibilities of where data analytics for corporate debt might be heading in the coming years.

We then include some closing remarks.

2

Corporate Debt Is Different

The Unique Nature of Corporate Debt

Fixed income is different from equities, and corporate debt is different from government, agency, and mortgage debt. In the corporate debt markets, you are analyzing not just the debt and interest rate features impacting the bond or loan being analyzed, but also the credit quality of the issuer and the unique structure of the specific piece of debt being explored.

The first widely followed analytical tool may have been the Dow Jones Industrial Average that was started in 1896 by Charles Dow and included 12 stocks; it was run by hand at the end of each day for many years. After evolving considerably, it may still be the most widely followed analytical investment tool.

In general, stock indexes are fairly widely followed and well known, such as the FTSE 100 of stocks from the London Stock Exchange, the Nikkei from Japan, the Standard & Poor's 500 (S&P 500). There are also a multitude of other lesser known and followed indexes too, including the Russell 2000, the Wilshire 5000, and the FTSE NASDAQ 500.

Data and analytics on stocks are widely followed and have fairly high functionality. However, building stock databases and related analytics and maintaining them, I would argue, is fairly easy compared with doing the same in the corporate debt market.

For example, most major companies have one class of common stock; however, the same company may have 10, 20, or even more different corporate debt instruments outstanding. Additionally, each debt instrument typically requires more information just to describe it than a share of stock and even more data to analyze it. A simple example of this is outlined in Exhibit 2.1 using the bond and stock data for Sirius XM Radio, Inc.

Exhibit 2.1 Example of basic details for debt and equity of Sirius XM Radio

BONDS (Amt. Out. In 000,000s)						
Issuer Name	**Coupon**	**Maturity**	**Amt. Out.**	**Rating**	**Ask Price**	**Yield**
Sirius XM Radio Inc.	4.250	5/15/2020	500	BB-	92.25	5.87%
Sirius XM Radio Inc.	4.625	5/15/2023	500	BB-	89.38	6.28%
Sirius XM Radio Inc.	5.250	8/15/2022	400	BB-	93.75	6.23%
Sirius XM Radio Inc.	5.750	8/1/2021	600	BB-	99.25	6.01%
XM Satellite Radio Inc.	7.625	11/1/2018	539	BB-	109.50	2.55%
STOCK (Shares Out. In 000,000s)						
Issuer Name	**Shares Out.**	**Price**	**Dividend**			
Sirius XM Radio Inc.	6,209	$3.85	N.A.			

This exhibit highlights a typical case in which a company has just one class of stock but five different bonds. Note that this does not even include any bank debt. The minimum information to describe the debt securities requires significantly more columns.

This complexity must be kept in mind when doing any analytics on corporate debt issues and especially when constructing databases for analytic use and using the data. You have to be careful that all of the necessary information to do various analytics is included in the database design and that the data is sortable and searchable by all the various categories of data in the system. The databases also have to have a flexible design so that new factors can be added into the database; for example, a new type of coupon structure that has not been seen before.

Let us look at some of the factors that make corporate debt so complex.

First, the basic description of a debt instrument involves more data than for a typical stock. The stock description usually includes the (1) name of the company, (2) shares outstanding, and (3) a price as well as a dividend rate if it pays one. However, the debt description needs (1) the company name, (2) the size of the bond issue, (3) the ranking of the bond, (4) the coupon, (5) the maturity, (6) the price, and (7) at least one measure of yield and usually a credit agency rating, too.

For a corporate debt instrument, there are a number of different types of yields you might want to look at—most notably yield to maturity, yield to worst, and current yield. Additionally, to calculate any yield on a bond, the system must have the data on any details about a call schedule, special calls, and sinking fund[1] data. It is also typical to show the spread of the corporate yield relative to a comparable maturity Treasury bond or comparable benchmark. You might want to look at duration and convexity measures as well. This data all focuses on the debt instrument and not on the quality of the company that has issued these instruments. Other noncorporate, fixed-income issues generally do not have the same level of complexity because they do not have the corporate aspects to follow.

Ultimately, corporate debt analysis should have at least three separate aspects that need to be utilized: first, the overall credit quality of the issuer; second, the structural analysis of the debt instrument that you are analyzing; third, the relative value that the specific debt instrument is offering you at the time when you are considering buying and selling it. From a larger market perspective, you also want information on how the market you are invested in is structured and weighted and what the critical statistics are for this market to be able to help in your overall conclusions. The more data and flexible analytics you have on all of this information the better.

Ideally, a database, especially one used for credit selection, should include some data for fundamental analysis as well. These files would include items such as the country in which the company operates, the country and currency in which the bonds or loans were issued (not always the same), and the industry in which the company operates. Then, there are basic credit ratios that you typically want to see, such as a leverage ratio,[2] an interest coverage ratio,[3] and perhaps some measure of free cash-flow generation. In addition, there might be any number of credit analytical items that would be beneficial to have access to; these might be similar to tools used in equity markets, such as equity market capitalization levels or equity trading multiples, but are likely to also include several debt specific analytics as well.

Keep in mind that all of this data just described does not even incorporate many details in the description of notes that are considerably harder to model. In a corporate bond prospectus for a new offering and in its indenture, it is not unusual for the "Description of Notes" section to be longer than the business description, especially when the company is a high-yield issuer. Some of the details that might be included could be bond puts,[4] equity clawbacks,[5] or a special 103 call. Then, there is typically a plethora of language describing the covenants. For leveraged bank loan agreements, the covenants often include maintenance tests—and these documents are even longer and more detailed than bond indentures. Detailed covenants can cover everything from reporting requirements, to the ability to issue debt, to what happens if the company is bought and by whom. These covenants are difficult to get into a database and are time consuming to analyze because the details can vary greatly from bond to bond (even among the same issuer). However, these covenants and terms can be critical in making an investment decision. All of these factors add to the complexity of corporate debt analysis.

Analytical tools for corporate debt instruments have some similarities to other major fixed-income instruments. All of these tools focus on much of the same "bond math" data: measures of yield and maturity and analysis of volatility relative to interest rate structures in the marketplace.

Government bond analysis obviously focuses on the stability of the underlying economy and is more focused on the general level of interest rates and the term structures. Asset-backed security analysis focuses on the cash-flow ability of the underlying assets and the details of the underlying issuing structure, among other things. Good corporate debt analysis not only incorporates the basic bond data and analysis focused on structure and interest rate sensitivity used by other fixed-income assets, but also includes all of the data about the underlying strengths and weaknesses of the corporate issuer of the debt and the structural aspects of the debt instruments.

This brief section highlights how many layers of data are necessary, not just in fixed-income data analytics, but particularly in corporate debt market analytics. The level of detail and accuracy of the data that you have and the relative flexibility that you have in being able to manipulate and analyze it can make a massive difference in the value of any analytical work you can do in the corporate debt markets.

Note that it is important to be able to do the top-down macro market analytics as well as the credit-specific, bottom-up analytics. There have definitely been cycles in the corporate debt markets where macro data and technical seem to overtake all of the investment decisions, and there are other cycles where it appears credit selection has been the only difference maker. The best practitioners always examine both through all cycles, and, no matter how much the macro data appear to overwhelm the market, you should never lose sight of the fact that ultimately the credit quality of what you own is paramount.

Sources of Data

You might define analytics as logical analysis based on taking—usually historical—data and looking at relationships between the data points.

So if all the analytics is going to be based on the data, then the quality of the data is often a major differentiator in the quality of the analysis. Remember that there are so many different data points that can be inputted on both an "issuer" of corporate debt as well as the actual debt "issue," that not only does the data in your database need to be accurate and updated, but the database you utilize also has to have all of the factors that you might want to utilize in analysis.

The sources where you can get data on corporate debt are not nearly as prevalent as they are for stocks or government bonds. Additionally, the size of the market for corporate debt, and especially below investment-grade debt (or high-yield debt), in both loans and bonds has ballooned in recent years in both the United States and internationally. So although there are more sources of data available, there is also more and more data to keep track of.

Remember, the data changes fairly often in the debt markets, too, bonds are called or tendered, and new securities come to market frequently. In some economic cycles, bond exchanges or bankruptcies happen regularly. Additionally, in the loan markets, amendments and changes in terms of the loans can happen often—these can occur with limited public disclosure. So updating the data even on existing debt instruments can be a major task.

Many of the available data sources come from large broker dealers/investment banking firms as well as third-party vendors, such as Interactive Data Corp., Bloomberg, and the major rating agencies like Fitch, Moody's, and Standard & Poor's. Most larger companies in the money management business, as well as on the investment banking side, develop internal proprietary systems or customize turnkey systems to track data themselves.

You might want to look at several sources to measure overall market performance as well. The most commonly used are indexes. The indexes are subsets of the entire universe of corporate debt. Well-constructed indexes can be customized and specialized for specific measurements. Most of these are currently developed and maintained by the investment banking side of the industry, and these widely used tools are covered in much more detail later in this book. Indexes are one of the best tools to use when trying to measure performance attribution and relative weightings. But if you are trying to do sorts and queries to find potential investment ideas, an index might not be helpful because it might not include all of the issues that you could invest in. For that purpose, you need to have an all-inclusive database, not just an index.

You can monitor the market's performance in other ways as well. You could look at the performance of public funds or mutual funds dedicated to a certain asset class. Morningstar and Lipper are two major sources for fund performance and closed-end funds are listed on the exchanges; these sources can also give information on flows in and out of an asset class. Also exchange-traded funds (ETFs) for corporate debt markets are newer investment vehicles that tend to use model portfolios to mimic some liquid index, and these can be followed as a measure of performance of more tradable securities within the corporate debt markets. Similarly, there are tradable indexes made up of selected liquid credit default swaps[6] that can be used to measure relative values on a spread basis as well as market trends. All of these nonindex products have some value as a tool to gauge market movements, but there are also numerous caveats attached to their use, which are discussed later in this book.

You might want to examine the performance of the market because you want to compare it with a portfolio that you are managing or you might want to compare the performance of a given asset class with other asset classes to decide where to allocate funds. This could be comparing performance or volatility of, for example, the S&P 500 with the performance of the investment-grade bond market, or perhaps the high-yield loan market historic yields compared to the U.S. Treasury rates.

Pricing Data

Although this chapter has highlighted some of the complexities of maintaining corporate bond data, the nature of the markets for these instruments actually makes accurate pricing data some of the most difficult data to get regularly. Although some efforts have improved pricing data for U.S. dollar corporate bonds, the data for the majority of bonds and especially for smaller issues, nondollar currencies, nonregistered bonds, and loans is still spotty at best.

Unlike most equity markets, the debt instruments are not typically listed on a public market, and many issues trade very infrequently. Even when bonds do trade, the prices might not be readily available except to the buyer and seller. However, there is now a semiregulatory source for bond trades that is readily available known as the Trade Reporting and Compliance Engine (TRACE) system.[7] This is not a real-time system. Additionally, many bonds are "Non-Trace," meaning their transactions are not reported on the system, usually because they are not fully registered, or known as 144A[8] for life. Additionally, the multitudes of non-U.S. dollar issues are not typically TRACE. Furthermore, TRACE does not give data on volumes either. However, the information is better than it was in the past, and it is expected to be adding 144A registered issues and may explore more detail on volumes. Even with these improvements, many bonds simply do not trade that often, so the pricing data is dated at best.

Other attempts to develop bond exchanges and trading platforms might give more pricing data in the future, but these seem focused on the most liquid bonds, on which there is already relatively good data.

Loans have no similar place where trades are recorded—and pricing data for loans is an even bigger issue than for bonds. The trading of loans tends to be less frequent, there are no regular listing requirements, and there is no required trade reporting. Note that, legally, loans are not considered securities. There are some attempts from newsletter services and pricing databases to report data on loan trades and pricing regularly, but it is generally spotty and focused on only the most active names. The data for pricing on loans is very dependent on broker dealer "runs," or listings of bids and offers, and it appears likely to be this way for some time.

Keep these issues about pricing data in mind as you read this book and do analytical work. So much of the important decisions are based on where you can buy and sell an investment, and, unfortunately, by the nature of this market, the data is limited and spotty in this market. As you do your work and reach a conclusion, be certain to revisit key pricing data to make sure that it was accurate enough to use and do not be afraid to question it. This illiquidity is often not properly analyzed or factored in by many market participants.

An Example

Analytics is used for numerous reasons and by numerous people with different purposes. Although it might seem obvious, one of the keys for successfully working with analytics and setting systems up is to know what the ultimate goal of the project is.

The analysis might be on a macro basis, such as trying to analyze how the net positions on a trading desk might shift with a major change in currency valuations, or perhaps on a micro basis, such as trying to analyze which bonds in a particular industry might be the least volatile.

Some typical analytic projects might be as follows:

- Performance measurement
- Performance attribution
- Portfolio/trading positions, weightings, and comparisons

- Industry sector, asset class, or country rotations
- Short-term, supply-and-demand analysis
- Credit ratings shifts
- Levels of interest rate risk

The following is a quick example of one project:

Assume that your mandate is to invest in corporate bonds that are rated between BB and AAA by the rating agencies. Your database tracks the trading level of the averages for each rating in this range, and it shows you that BBB-rated bonds are currently trading at their widest spread to A-rated bonds in the past five years and BB bonds are trading fairly tight to BBB-rated bonds versus historical averages of the last five years.

To judge these relationships, you might use relatively simple graphs and tabular analysis. A simple graph of spreads over the last ten years is shown in Exhibit 2.2. Examining the chart would lead you to conclude that BBB names are cheap and you should increase your weighting in this category of bonds.

Of course, you would also want to consider other factors to make sure there is not some other reason for these trading relationships (perhaps there is a coming glut of issuance of BBB-rated bonds or perhaps some major issuers in the BBB-rated category look like they will be downgraded to a lower rating category soon).

However, assume you have used your macro analytics and decided that you want to add BBB bonds. Now if you have access to a good database, you can use query techniques to find the bonds you would want to add. If the query tools are reasonably good, you can search the databases for BBB-rated bonds and then add queries for sorting them by the widest spread to the tightest spread and then add other queries that might meet your investment needs. These other layers of queries might include size of issue, selected industries, and perhaps certain financial ratios. What you are able to search for and sort, as well as what securities are included in your outcomes, depends on how robust, inclusive, and flexible your database is.

Exhibit 2.2 Illustration of spreads of various rating categories

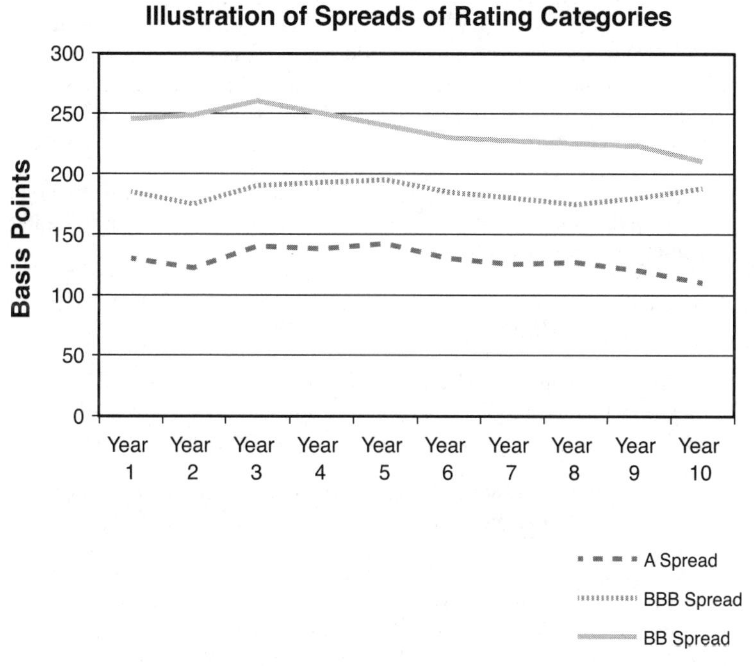

This simple example starts with some macro relative value analysis of certain tiers of the market, primarily using simple graphic and tabular methods to see what types of relationships there are in certain subsets of the market, and then moves to more credit-specific micro tools using database queries and sorts to find specific investment opportunities that fit the macro theme. Then, analysts can review those lists and prepare relative value analysis to decide which specific bonds look the most attractive and fit the risk-reward scenario of your investment assets.

Endnotes

1. A *sinking fund* is a requirement within the covenants of a bond that requires the company to retire a certain portion of the bonds prior to maturity; it is not very common any more in corporate bonds.

2. *Leverage* in this case refers to the amount of debt on a company; in Europe, it is often referred to as *gearing*. The *leverage ratio* examines the amount of debt relative to the cash flow the company generates. The cash flow is typically measured by EBITDA, or earnings before interest, taxes, depreciation, and amortization. EBITDA is usually adjusted for other noncash items on the income statement, which is usually referred to as *adjusted EBITDA*. However, this text just uses the term EBITDA.

3. *Interest coverage ratios* typically use EBITDA and divide it by the total interest expense of the company. This is a very rough measure of a company's liquidity or the ability to service its debt.

4. A *put* in this case is an option of the holder of a security to sell it back to the issuer at a specific price.

5. A *clawback* is a feature that has evolved in high-yield bonds that allows the company to buy back a portion of the bonds at a set price (usually par plus the coupon) if it raises equity money or undertakes an initial public offering of stock.

6. *CDS* is a credit default swap, which is a form of default insurance (or option) that is used in the debt markets and explained more fully later in the book.

7. *TRACE* is a FINRA-developed vehicle that facilitates the mandatory reporting of over-the-counter secondary market transactions in eligible, fixed-income securities. All broker/dealers who are FINRA member firms have an obligation to report transactions in corporate bonds to TRACE under a Securities and Exchange Commission (SEC)-approved set of rules.

8. *144A* is a form of private registration of bonds that is widely used in the U.S. debt markets; the bonds are limited to Qualified Institutional Buyers. Often, the bonds are sold with registration rights, which means they can become fully registered bonds over some time frame, usually within a year.

3

Managing Projects and Managing People

The Basics

This section could probably just include the following declarative sentences:

- Target your goal for the project and communicate it.
- Communicate thoroughly with everyone on the project.
- Document the process.
- Discuss the results, decide on an action plan based on the results, and communicate it.

Notice that the word *communicate* is the most used word in that short list, and the lack of communication often makes these projects very inefficient.

In most organizations that work with corporate debt, internal systems are used to run and maintain at least some of the data, whether they are purchased from third parties and customized or self-developed. The people who are the primary end users of the data and analysis are usually not the people who are managing, running, maintaining, and programming the data and are often not the people who are doing the analysis of the data.

Most projects include the ultimate end user and a technology or systems professional. In many projects, there may be an analyst who sits in the middle of the process who is going to make an initial conclusion on the data and may ask for multiple iterations of the data before passing it on to the final end user with an opinion about what the data shows.

It is critical to the successful use of data analytics that all of the parties communicate. The different groups of people involved in this process often have very different types of backgrounds and training. It is not uncommon for one group to have a background in finance and the other in computer programming. The people have to understand each other's jobs and to some extent each other's business "language."

The end users of the analytics could be in any number of jobs. These could include investment bankers or capital markets people, credit analysts, strategists, traders, or salesmen. They could also be risk managers or trading managers who want to monitor various aspects of a business. They could also be money managers, including portfolio managers or analysts, or often they are fund marketers. They could also be institutional investment advisors or asset allocators for wealthy individuals, pension funds, or endowments. Obviously, this broad-reaching group of people could be doing very different types of analytical projects. Most of the people in these roles will have a background in finance of some sort.

The people who are running the projects, coordinating the output from the database, and setting up the databases and interfaces to be used for these projects all tend to have a technology, mathematical, or some other quantitative background. Sometimes, the background is in programming or systems management, but it is rarely a pure business and finance background.

Communications

One of the major items decreasing the efficiency of most projects is that the scope of the communications between the team members is too narrow. For example, the end user often comes in with a "one-off" request; perhaps a portfolio manager comes to work and, in reality, is worried about rising interest rates, but all he asks the analytical or programming team is, "Can you get me a list of the largest bonds in the market listed by duration[1] and maturity?" And then this assignment is dutifully fulfilled. Often what happens is the end user gets too much information to use efficiently and needs it cut down in different ways, perhaps by issue size or by ratings. He then might want a bunch of different sorts and might want to back test his theory that this is the right way to be positioned for a defensive posture during a rate increase.

It is much more efficient if there is a dialogue and the end user gives the complete picture saying, "I am worried about rising interest rates. I assume the best thing to do is to look for bonds with a short duration to buy. What are the best ways for me to look at this?" Hopefully, the analyst or the programmer then points out that he could print out a list of short duration bonds, but he might want to limit the list by the rating category that fits the portfolio manager's investment style. The analyst might also point out that he could also track the last two or three times there was a sharp rise in general interest rates and look at which subcategories of the market, other than short duration, performed the best and the worst during those cycles and then do a sort of them. He might also suggest several different types of duration to run in some sorts of the data (there are several different ways to calculate duration), such as using option-adjusted spreads and duration.

The end user of the data also needs to be careful how he or she asks for data, so that the requests are as efficient as possible. In the heat of changing markets, this is not always done. End users should also have a realistic understanding of what is involved in undertaking these projects to obtain data they are looking for and the amount of time that data requests might take. This is integral to help prioritize work and have realistic time frame expectations.

The people who are designing the systems and programs have to understand that sometimes if a project takes too long it is useless because the markets have moved on. So they need to be honest and communicate the time frames and manpower needed for projects, while keeping an eye out that they do not become a bloated cost center. The problem at the other end of the spectrum is that they fall into becoming the "no, we can't do that" center.

I once read about a U.S. FBI agent who was being quoted. The agent effectively said: Give me enough resources and I will catch anybody. I think good systems managers should have the same reasonable arrogance about their abilities—but they need to be pragmatic because in the rush of day-to-day business, resources are not always allocated for these valuable tools, but the answers are always needed. Therefore, they need to manage their resources carefully, they need to have an organized workflow for their team and for third-party vendors, and they need to understand that the workflow will likely be in constant flux, just as the markets they are analyzing are in constant flux.

Systems have to be designed to be user friendly. Ideally, end users should be able to look up much of the data themselves. A subtle danger that can slip into the work of the systems team is that they start to make their operations so complex that they make themselves indispensable, meanwhile building a poorly designed system. A program or system that is not designed for others to be able to interface with easily is simply a bad system.

It is tricky because the systems designed to manipulate and use the data have to be adaptive, new debt structures and data categories need to be added, and the system has to be able to handle them. Additionally, special projects arise that need to be addressed. One suggestion in managing these requests is simply to create a prioritized wish list of the forward-looking projects to be addressed when there are slowdowns in daily requests or there is the ability to add resources.

Both sides have to understand the full extent of the problem that is being addressed and develop realistic expectations of timeliness—and both of these need to be communicated.

As mentioned, the goal of every analytical request should be defined and there should also be a plan of action for every request. This should include a review of the data and a decision about any actions that should be taken. If actions are going to be taken, it should be clear what those actions are and a time frame to undertake these actions should be identified. The conclusion may be no action, a detailed trading or management strategy, or to do more analytical work (of course). The conclusion about what action will be taken should be shared with the team that worked on the project; it will help them in understanding what their work is being used for and increase their engagement. The work should not ever fall into a vacuum without any conclusion.

Documentation

Documentation helps to make everything more efficient. A written record of data and analytical requests and completed tasks is a valuable reference tool. Ideally, you can include what the project is and the goal of the project. If the request is kept in a folder with a type of reference system, it can be very helpful when there are similar requests in the future.

The data and systems people should also keep documentation of the steps they are taking for the project. This accomplishes several things: (1) As these processes are often iterative, it helps to keep track of each subtle change that may have been made in each round of the work, so you do not end up redoing things; (2) it helps give a road map to anyone else on your team who might need to either take up the project for you or revisit it down the road; and (3) it gives you an outline to write up what was done for the project when you inevitably have to review the project with other end users. These documents should all be stored in the same file related to the project. *Note that this documentation should not become a project within itself.* The notes should be kept as simple as possible and, while being exacting in their language so misunderstandings are minimal, the notes can be as simple as e-mails saved to the proper files. Detailed titles for each file will be helpful in this regard as well.

The key here is that everyone involved should understand as much as possible about the overall goal of any projects and what happens to them. Being inclusive in strategic and business meetings can help each group understand the process and goals involved. This should help generate new ideas and better processes and increase efficiency.

Endnote

1. Duration is a measure of interest rate sensitivity.

Closing Comments on Section I

The corporate debt markets have many layers of information and some glaring gaps in data, which make analytics challenging. From the number of debt issues outstanding, to the level of detail required for the description of the debt instrument, to the various measures of yield and the underlying credit metrics, there are numerous data points needed to analyze these markets. Additionally, there is spotty pricing data. All of this combines to add significant complexity in using and developing analytics in the market.

Most of these degrees of difficulty are even greater in the bank loan markets. Disclosures vary greatly for each transaction and terms can change on loans (with the approval of holders) with little or no public notice. Fortunately, loans usually benefit from being the most senior debt in the capital structure.

It is important to remember one of the major issues in using analytics in this marketplace is the quality of the data, so that the analytics can be relied on. Try to have as clear an understanding as possible when you have to be wary about the data. Always question any conclusions and don't be afraid to doubt.

At the end of the day, successful investing consists of owning the right securities. To this end, data analytics in the corporate debt and loan markets is not a final answer, but is an important part of the equation for strategic decisions. The unique nature of each issue and issuer requires a detailed, bottom-up approach as well as the top-down approach used in data analytics. Credit work and credit selection at the right price are critical. Keep in mind that in fixed income, and especially in the leveraged markets, long-term success is often predicated not just on owning the right securities, but also on making sure you do not own the wrong ones; if analytics can help weed out the latter, it has proven very useful.

If a modern-day practitioner in the securities industry does not use data and analytics, he or she will likely fail; however, if this is all the practitioner uses, he or she will likely fail, too.

Section II
Terminology and Basic Tools

By Robert S. Kricheff

The first chapter in this section reviews some of the terms that are used in the debt markets and some of the protocols; it takes you through some examples and discusses the advantages and disadvantages of some of the terms. For those experienced in the markets, this chapter may be superfluous.

The second chapter in this section outlines some of the typical tools that are used in analytics in the market. Many of these are in the family of statistics and will be familiar to many of you. This chapter provides some simple examples and then later chapters show you how these tools are used.

The final chapter in this section is a brief summary of data mining. This is an analytical tool—or actually a series of tools. This topic warrants an entire chapter dedicated to it because it is a significantly newer field and has been getting significant publicity in recent years. It is not currently heavily used in corporate debt market analytics but will likely increase in its usage in coming years.

4

Terms

This chapter reviews some terms and their protocols as used in the corporate debt markets and in analytics.

Valuation Terms

The following are some common terms used in valuing corporate debt and brief descriptions of how they are used.

Prices on corporate bonds are quoted as a percentage of face amounts. A typical bond is $1,000 or €1,000 face amount (or par). Even though in practice, prices often appear with a dollar sign or a euro sign in front of the price, this is incorrect and it is actually a percentage of par. Bonds typically have a minimum size of $1,000. So a price of $91 actually means 91% of $1,000. A *round lot* trade is usually considered to be $1,000,000 face amount (or one million in whichever currency the bond is issued in). So, using the same example, a round lot of $1,000,000 at 91 would actually cost $910,000. Typically, a person would say, "I have bought one million bonds at 91."

The *yield to maturity* (YTM) is the calculation of the return that a bond will give you if you buy it at a given price and it pays off at maturity. When a bond or a loan is callable and the bond is trading at a premium price (a price above 100), the same price might give you different yields depending on which date the bond might be called, or if it goes to maturity. The *yield to worst* (YTW) is the lowest yield of these various possibilities and is the one that is commonly used to be conservative.

The *spread to worst* is the difference between the YTW and a riskless bond of a similar maturity. So if a bond is trading to a yield to worst of 7% to a call date in three years and the three-year Treasury is trading at 2%, the spread is 5%. However, the spread is usually quoted in basis points (bps). A basis point is equal to 1/100th of a percentage point—that is, 100 basis points (bps) per 1 percentage point—so the spread to worst in this case would be quoted as 500bps. Ideally, the riskless bond you are using to "spread" the bond off of should be in the same currency as the bond.

There are other types of yields and spreads, such as current yield, option adjusted spread (OAS), and others. However, YTW and STW are the two most commonly used. *Current yield* is simply the coupon divided by the price.

When and How to Use Yield and Spread

Yield and spread to worst are commonly used when using corporate bond data and running analytics on relative value. There are arguments for which you should use and I believe there are different times when one should be used versus another.

Investment-grade bonds are typically quoted by spread, as they tend to trade fairly tight to treasury rates and more in unison with treasuries. Additionally, spread helps to normalize relative value of bonds over different maturities because it is adjusted for the yield curve by using a maturity of a government bond that matches with the bond you are analyzing. Investment-grade maturities tend to vary more than maturities in high-yield bonds and in leveraged loans. In high-yield bonds, typically the YTW is used because yields can vary so greatly and drive so much of the return, but the STW is key when comparing specific credits and relative value of differing maturities.

In data analysis, you may compare spread to worst of different markets over time (such as the investment-grade and high-yield markets), various sectors of the market over time (such as the average for Senior Secured Notes versus Senior Notes), or specific bonds within a subsector (such as all of the bonds in the Energy sector) along with numerous other comparisons. Although spread works well in all of these cases, never completely ignore looking at yields. For example, bonds trading to a very short maturity or call date may offer fairly high spreads, but because shorter dated treasuries are generally much lower yielding, these bonds may end up with meaningfully lower yields than other bonds you are looking at, which can impact returns significantly. If you load a portfolio with only short-maturity, high-spread, but low-yield bonds you will get very skewed returns.

Bank Loans and Yields

Bank loans present another handful of problems as to what yield to use, they are often callable immediately or within a year, and they also typically have floating-rate coupons, which change every three or six months. Today, most leveraged loans also have a *LIBOR floor*. LIBOR is usually the base rate off of which a loan's interest rate is based. A floor is a minimum, which this base rate is. For example, if the LIBOR floor is 100bps and the current LIBOR rate is 30bps, the loan's interest rate would be based off of 100bps because that is the higher number. Typically, the interest rate on a loan would be quoted like "L+375 with a 100bps floor," which means that the loan will pay an interest rate of LIBOR plus 375bps, but if LIBOR is below 100bps, it will pay an interest rate of 475bps. LIBOR is the London Interbank Overnight Rate, a short-term lending rate to banks.

There are several different ways to run yields on loans. Some of the common ways include the simple current yield, which is the interest rate divided by the price. Another way is to run it based on the implied LIBOR curve through the life of the loan. Yet another is to assume the cost to swap the bond into a fixed-rate instrument and then run the yield like a bond. More recently, people have run a yield assuming a take-out in three years and utilized the current interest rate. The discount margin is often used as a comparable measure to a bond spread. Which one you need to use depends on what you hope to do with the yield. It can sometimes seem like figuring out the correct measure of yield on bank loans can be like trying to measure clouds, but the key to remember is that in longer-term studies or comparative analytics, it is important to be consistent in which measure you use.

Volatility Terms

Measures of volatility are commonly used in analytics for investments. Some measures are used on a relative basis and some on an absolute basis. Volatility of returns is often used as a measure of risk. Let us start by going over a simple definition of total return and then examining certain measures of volatility.

Total return is the measurement over a given period of time that would measure the interest income and the price movements as a percentage of the price that was paid for the bond at the beginning of the period. Total return measurements are used to compare asset classes, subsectors of a market, and individual bonds and loans.

Standard deviation[1] is often used to examine the volatility of returns. Investors often want to examine the total return of an investment relative to its volatility over time. Volatility is generally viewed as risk. Measuring the risk versus return of investments is very important for all security analysis. One of the most common tools for this is the Sharpe ratio, developed by William Sharpe. It measures the adjusted return of an asset, or a portfolio, divided by its standard deviation. The adjusted return takes the return of the asset and subtracts the return of a theoretically "risk-free" asset; usually U.S. Treasury bonds are used as the equivalent for the risk-free return in U.S. dollar denominated securities.

Beta is a measure of volatility in the return of a security relative to a benchmark. (The benchmark is the proxy for the market.) A beta is run from a regression, and a beta of one means it will move in unison with the market; higher than one indicates it is more volatile than the market; less than one is less volatile. Beta (and alpha, explained next) are both part of the *Capital Asset Pricing Model* (CAPM), which is a finance theory that factors in the time value of money (using something like a U.S. Treasury curve), the assets risk (as measured by beta), and expected return. This calculation is weighed versus a benchmark to determine whether the asset should be purchased or not. The CAPM and alpha and beta can be applied to an individual asset (like a loan, stock, or bond) or a portfolio.

Alpha takes the volatility in the return of an asset relative to a risk-adjusted return of a benchmark. For example, the return of the benchmark is adjusted for the beta of the asset and then any return over that can be referred to as the alpha, sometimes viewed as the credit selection in a portfolio versus the overall impact from the market.

Alpha and beta might not be the most common terms you will run across when undertaking corporate debt analytics. However, they are important and tend to be used significantly more when doing work for money managers or asset allocators in determining performance and performance attribution of portfolios or asset classes. Along with standard deviation, these are all measures that help to measure return versus risk and are often used in concert with CAPM analysis.

Duration is another measure of volatility but is unique to fixed-income instruments. Theoretically, this measures the volatility relative to a change in interest rates. Duration takes into account the price paid for the debt, the interest rate, and the maturity of the bond and measures the volatility in the price of the security relative to any change in interest rates, all other things being equal. Lower coupon, longer maturity issues tend to have higher duration and are more volatile to these changes.

Further Discussion on Duration

Although classic analysis states that duration is related to a change in interest rates, keep in mind that, especially within the leveraged markets, duration can help to measure a bond's response to a news event that impacts credit quality and affects what interest rate the bond should trade at.

For example, say there are two bonds issued by the same company:

- Bond A Senior Notes that have a maturity of five years and an 8% coupon
- Bond B Senior Notes that have a maturity of seven years and a 7% coupon (this can happen because they were issued at different times)

If the company reports very bad earnings, Bond B will see more price movement than Bond A because of the longer duration.

Another item to keep in mind about duration is extension risk. This is particularly true in high-yield bonds because most are issued with call structures, whereas many investment grade bonds are non-call for life (also known as *bullets*). If a bond is trading at a high enough premium, it may be trading to an early call date rather than the maturity date, which might help it to show fairly short duration. However, if the price drops, it may suddenly "extend" out to a longer date and then become a higher-duration bond because of its negative convexity.[2]

In analysis of trading positions, portfolios, performance attribution, and individual bond analysis, duration is often one of the key components that is analyzed. Historically, in the leveraged markets there has been less attention paid to duration because credit gains and declines and corporate actions tend to outweigh the impact of interest rate movements during most market cycles; however, it comes dramatically into play when this sector is trading at a tight spread or when interest rate movements become volatile.

Debt-Ranking Terms

Rankings of bonds are critical to understand in debt analysis. Bonds have different ranking descriptions. Typically these include the following:

- Senior Secured
- Senior
- Senior Subordinated
- Subordinated

There can be nuances in these rankings, of course. Bonds can be ranked as Secured, but there might be bonds such as a Senior Secured First Lien and a Senior Secured Second Lien, or there could be a Subordinated and a Junior Subordinated Note, but the categories in the preceding list are the major rankings. If a bond is listed just as Senior, it is usually Senior Unsecured. If no ranking is stated, bonds are typically Senior Unsecured. The legal documents that define the terms of a bond or loan (the indenture or loan agreement) give detailed information about the ranking. If the bond or loan is secured, there will also be a detailed collateral agreement that outlines the terms of the security.

These rankings determine which tranche of debt would get paid off first in the event of a bankruptcy. Corporate debt analysis often slices and dices various data by these tiers and looks at the relationships between the various ranks of bonds. This type of tiering of the market and of individual capital structures is often used in relative value analysis within the debt markets.

Corporate Structure and Bond Rankings

Like so much in the corporate debt market, you must often be aware of exceptions—or major mistakes might be made. One of the most common mistakes within this area is not differentiating between operating company (OpCo) and holding company (HoldCo) bonds. In many corporate structures, an OpCo has the first claim on the assets of the company, but then, for any number of reasons, another company is formed that is once removed from the assets and simply owns the stock of the OpCo. In this case, let's assume all of the assets of the company are at the OpCo. As far as getting paid off by the assets of the company, no matter what the ranking of the debt at the OpCo, it ranks ahead of the debt at the HoldCo in a bankruptcy as far as its claims on the company assets at the OpCo because the HoldCo debt is "structurally subordinate." Now here is a twist: Holding company debt is often issued as Senior or Senior Secured debt. There are periods in the markets (particularly when the market is strong and private equity-owned companies are paying dividends) when holding company bonds get issued in large quantity. If you run an analysis of the spread between senior secured and subordinated bonds historically and do not separate out HoldCo Senior Secured Notes, you will get a very distorted picture of the risk premium that the market is requiring between the two rankings. There are also sometimes unusual exceptions such as where a note and a loan may be ranked equally, but within the intercreditor agreement, the loan has a priority in a bankruptcy. Sometimes understanding debt rankings and corporate structures can be like peeling layers of an onion, with pages of legal writing on each layer, but you should be aware of these types of nuances and be cautious anytime something looks unusual or too good or too bad.

In Exhibit 4.1, the holding company notes are structurally subordinated to the operating company notes, even though they are called Senior Secured because all the asset value is at the operating company, and the holding company notes are only secured by the stock of the operating company. You may ask then what the HoldCo notes are secured by. Typically the answer is the stock of the operating company.

Exhibit 4.1 Holding company and operating company structure

Example of a Holding, Operating Company Structure
Holding Company
9% Senior Secured Notes
Operating Company
7% Senior Notes
8% Senior Subordinated Notes

Credit Ratings Terms and Usage

Credit ratings are supplied by a number of independent agencies around the globe, but by far the most widely followed are Moody's and Standard and Poor's (S&P). The credit ratings are supposed to measure credit risk, and to a certain extent, they do. However, many experienced people in the market, particularly in the leveraged markets, view the ratings as very loose guideposts, fairly backward looking, slow to change, and of limited use in truly measuring risk and certainly massively limited in measuring value.

Changes in certain ratings can impact how a bond trades in the leveraged markets and in the investment grade market. These ratings changes have an impact because of the rules that investors impose upon themselves about which types of bonds they can own. Therefore, certain ratings changes will either open up or close down the number of potential buyers for a security or loan. This will cause some of these key breakpoints to impact the trading levels. Some key breakpoints include a move from below investment grade to investment grade, which typically elicits a rise in security prices, as new buyers can buy the bonds (BBB- and Baa3 are the lowest investment-grade rating from the two major agencies). Also a move down to CCC+ or Caa1, the highest of the "triple C" ratings, often limits the number of buyers for a debt instrument. Typically, once a credit is investment grade, its trading levels will be more dependent on its rating than in the leveraged markets.

Regardless of how market participants view the value of the ratings and their various shortcomings, they do offer a structure for a very general analysis of credit risk by issuer and, therefore, much analysis on a macro basis in the corporate debt markets will be run based on relationships between the rating categories.

Remember that many markets are defined by their ratings categories, so there is also often analysis about how "fallen angels" (bonds that have been downgraded from high grade to below investment grade) and "rising stars" (bonds and loans that have been upgraded from below investment grade to investment grade) have changed the overall composition and size of the various markets. It is also common from a macro basis to analyze ratings drift, which explores if a market or portfolio is adding risk or subtracting risk. In some cycles and in some analysis, people look at ratings shifts as a statement on the health of the economy as well.

Industry Group Terms and Definitions

Analysis by industry group is generally more associated with the equity markets than in the traditional, fixed-income markets; however, in the corporate debt markets, and especially in the more leveraged segments of the market, industry groups are extensively analyzed. Traders are often divided by the industry sectors they trade, versus in the treasury market where they are more often divided by the maturities that they trade.

Industry groups are fairly self-explanatory and analysis is commonly done using various industry categories for strategic analysis, attribution analysis, and credit idea generation, as well as relative value. However, it is important to note how they are defined when doing comparative analysis of various industry sectors. There may be major differences in how a portfolio manager may categorize certain issuers, by industry, and how some benchmark index is defining the same sector. For example, assuming there is no separate satellite sector and there are a number of issuers with large capital structures in the satellite industry that supply services to cable operators, broadcasters, and telecommunications companies, which of these three industry categories should they be categorized in? Two experienced market participants may disagree and it can impact how you analyze the industry data. So keep this in mind when using the analytical tools on industry categories.

Endnotes

1. *Standard deviation* is a measure of dispersion of the data. To understand why this is used, first you need to understand dispersion, which is typically calculated using the difference between the mean of the data and each data point, using the absolute value and then dividing it by the number of data points; the variance squares each of these data points before dividing them. The standard deviation takes these squares and runs the square root of them to create a measure of dispersion more comparable with the units being measured.

2. Convexity measures the curvature of the relationship between bond prices and yields. It shows how the duration (which is a linear measure) changes for a bond as prices and/or interest rates change. It comes into play considerably when bonds are trading at a premium and when they are callable; sometimes this aspect in the bond market is referred to as *extension risk*—as a price of a bond declines, it shifts to trading to a different YTW date (that is, call date) and that "extends" the bond's maturity.

5

Basic Tools

Introduction

Most of the tools and techniques that are currently being used for data analytics can be run using Microsoft Excel and similar programs. You will see that most of the widely used techniques in corporate data analysis at this time are fairly simple, though more complex methods have been creeping in. The math and the terminology used in corporate debt analysis can seem a bit daunting if you are not familiar with it, but the actual formulas and calculations are not nearly as important as understanding how to read the output of these calculations and what this output is, and is not, showing you.

There are programs and systems that run many of the bond calculations and can manage a large number of positions that also download pricing data in the corporate debt markets. The most widely used is probably the Bloomberg system. However, there are other vendors that have portfolio and trading position systems that are used as well and some firms have their own proprietary systems.

Within the world of the day-to-day finance and securities business, as well as theoretical financial analysis, there are numerous analytical and statistical techniques used. We do not pretend to address all of them, but try to focus on some of the basic ones used most commonly in the corporate debt markets.

Graphical Data

Laying out data in graphical form is relatively simple and can often be a great help in catching trends and changes in relationships between subsets of the markets or individual debt instruments and the markets.

Graphical presentation can be particularly helpful when looking at many of the macro trends impacting the markets. A simple example of this might be a time series that tracks the movement in the ten-year U.S. Treasury bond and the average yield on single B-rated corporate bonds as measured by some index. If you added in BB-rated and CCC-rated bonds to this graph, you might start to see a pattern of which type of bonds react the least and the most to interest rate increases (this of course is using the ten-year U.S. Treasury bond as a proxy for general interest rates). When you graph this data, you might also see that at a certain time, some of these rating tiers are trading particularly cheap or rich relative to other tiers versus historic relationships. This might trigger more work to try to generate some trade ideas to take advantage of this relationship.

A Caution about Historical Data

As with all of these analytical techniques, you have to be cautious about nonobvious items that might be influencing the data you are looking at. For example, a major caution I would give about exploring time-series data, as in the preceding example, is that the components in any segment of the market can change meaningfully over time and this can significantly skew the data or cause an unintended bias in it. This can be especially true within leveraged finance, as the components and structures in the leveraged loan, emerging market, and high-yield bond markets change more frequently than you typically see in the investment-grade bond market. For example, it may appear that the telecommunications sector of the high-yield market has tightened considerably to the rest of the market during the last dozen years, so you might conclude that this segment of the market is trading too rich, or you might conclude that the management of the companies that issue bonds in the high-yield telecom sector has done a great job of improving their credits over this time. In fact, what may have happened is that the issuers in that sector have completely changed during that time. At the beginning of the time period, the sector may have been dominated by bond issues made up of start-ups and proceeds were used for construction funding; many of these may have gone bankrupt and then over time a whole new list of better-established, cash-flow producing companies issued debt and completely changed the makeup of this sector. So, in fact, comparing the sector over that dozen years may be like comparing grapes and wine—there are similarities, but they certainly are not the same products and have a different effect when you put each in your body (or a portfolio).

Trend Lines

When graphing data, it can also be helpful to include trend lines. This can be useful, particularly when some of the data might appear more random when first graphed. The trend line may allow you to see more clearly the direction of the data and identify outliers as well. This is particularly true relative to a scatter diagram. Most programs such as Microsoft Excel have functions that can add in trend lines for data. A somewhat similar, but more advanced, technique is regression.

In Exhibit 5.1, the first graph shows a scatter diagram of bonds using their ratio of spread/leverage. The higher the company is and the closer to zero on the X axis, the better. The first graph looks fairly random. When the trend line is added, as in the second graph in the exhibit, one can see that some data points that lie above the trend line may look cheap relative to this universe and those that lie below the line may look rich. Theoretically, the trend line here implies the fair value target for each example.

Exhibit 5.1 Illustration of scatter data with and without a trend line

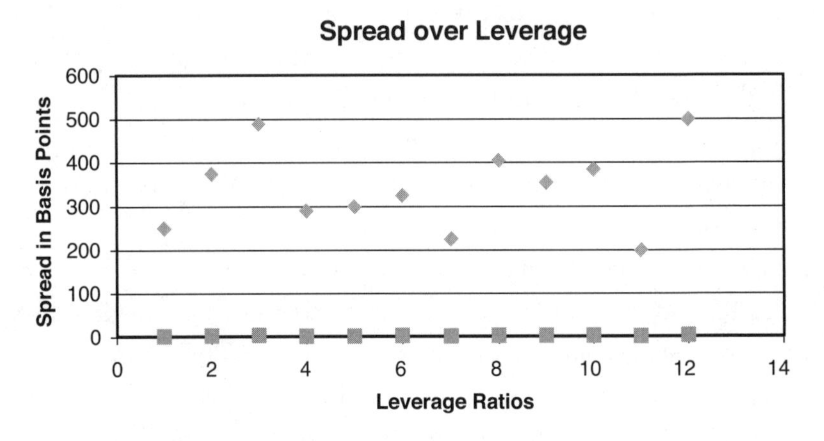

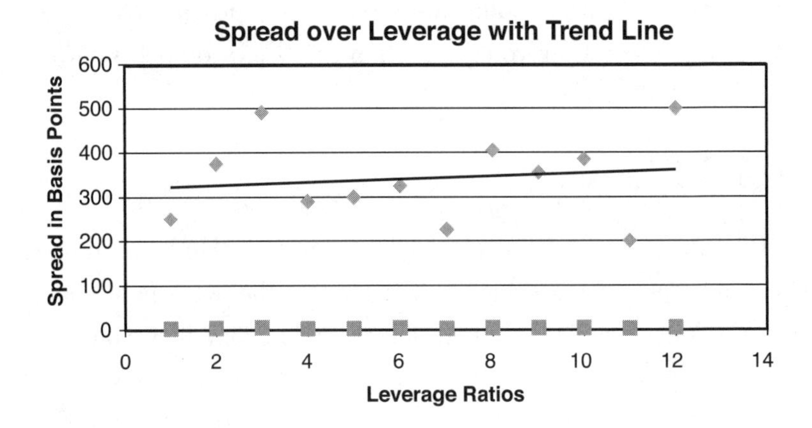

Regression

Regression analysis effectively takes a dependent variable (let us say spread to treasury on a group of bonds in the same industry), uses one or more independent variables (for example, a leverage ratio and/ or interest coverage ratio), and analyzes the amount of influence the independent variable or variables have on the dependent variable. Certain mathematical tests can then be run to see if the independent variables can be used as a predictive tool for the dependent variable. In plainer English, the model hopes to tell you if the independent variables could be used to predict the dependent variable. (For example, in this example, the question is: Do the leverage ratio and the interest coverage ratio have enough impact on the spread of a bond to use these ratios as a predictive tool as to where a bond should trade on a spread basis?) You often see simple, single-variable regressions used in credit analysis for cheap/rich analysis of comparable bonds. The line that is generated from regression has a formula that gives you coefficients for each independent variable used in the formula.

This text is not positioned to go into a detailed discussion of regression analysis, but there are a plethora of books that do. Microsoft Excel and many other programs can easily do a very simple regression analysis, which can be very helpful. However, the ease with which these programs can be used can lead people to use regression analysis almost too often, without checking on how correct it might actually be to use. However, you should be aware of a few cautions on using regression analysis. There are important measures of how well the data fits to the predicted line or how much of the dependent variables can be attributed to the independent variables. The coefficient of determination, or r^2,[1] is the main measure of how well the change in the dependent variable is attributed to the independent variable. When doing a regression, you need to be sensitive to using the work if this measure is too low—generally above 0.6 is considered statistically significant. Some other cautions follow:

- Remember, the regression is based on past data and over time markets change (corporate debt markets can change dramatically, both in what the components of the market are and how those components are structured), so be cautious of what the real predictive powers of regression are if there have been major changes over time within the markets you are testing.

- Simple regression works if the relationships between the dependent and independent variables are linear; many relationships are nonlinear. A simple example is if you are measuring weekly wages per hour, and the hourly wage goes up for overtime, the relationship between hours worked and weekly wages will not have a linear relationship during any period when there is overtime.
- One common technique to "smooth" the data is taking the logarithm (log) or the anti-log of the data and then regressing; there are many other techniques to smooth data as well, but be sure these techniques are used consistently throughout a study.
- Keep in mind that regression tends to work best when there is a large sample size; the smaller the sample size, typically the less predictive or reliable the analysis can be.
- Another technical point often seen in attempts at multivariable regressions in the debt markets is *colinearity*. If the components being used as the independent inputs are themselves correlated (having a high r^2), it can result in misleadingly strong correlations—there are tests for this.

There are several other tests that should be run as well to be truly comfortable with regression analysis.

The concluding point here is that regression can be a very valuable tool, but be careful that you use it properly and do not overuse it or become overly dependent on it. As mentioned, there are many excellent books on statistics and econometrics that delve considerably more into the process of regression. Excel can run regressions, but there are also much more powerful statistical packages such as the open source R programs and ones that can be purchased such as the miniTab and SAS programs that can be used to run data.

Correlation

In statistics, there is the coefficient of correlation, or r, and the coefficient of determination, r^2 (the latter was discussed previously). Correlations, using the r coefficient as the measure, can be used very independently from a regression. Correlations are often used to measure the strength of a relationship between several asset classes. Correlations in corporate debt analysis are frequently run comparing total returns for various sectors of the investment world. The r coefficient is in a range between –1 to +1, with these boundaries equaling a perfect positive or negative correlation. Often, you might want to take outliers out of the tests as well; I often ignore the 2.15% of the samples at each extreme, mimicking the portion of data that would be two standard deviations away from the mean in a normal distribution.

You often see studies that show the correlation between various asset classes and some benchmark, such as the ten-year U.S. Treasury or the S&P 500 compared with the investment-grade corporate bond market and the high-yield corporate bond market. When measuring returns over time, these studies have often shown that investment-grade bond markets are much more highly correlated with treasuries, whereas high-yield bond markets are much more correlated with the stock index. But correlations can be run in an endless number of combinations; they could be run by industry subsectors versus an index or versus each other, or by credit rating category or maturity. Remember, though, this analysis shows that certain sectors or asset classes have moved in unison or not; it does not prove any cause, dependence, or direct impact of one on the other (or lack thereof).[2]

Backing Up Graphs with the Data in Tabular Form

Generally, when graphs are formed from data, the data has been put in tabular format. Of course, eyeballing the data in tabular form is very helpful, too. When putting data into tables, try to have the data be as detailed and as maneuverable as possible. You will likely want to be able to sort and search this data, often repeatedly with varying criteria.

There are also many times when laying out the data in a simple, tabular format is the best way to examine the information; color-coding or some other type of highlighting might be added to enhance the visibility of certain trends or outliers (perhaps as simple as red for negative numbers or a specific color for more than a 10% deviation from the mean). Remember, often a simple table of data can be the best way to show something. As Albert Einstein is quoted as saying, "Keep everything as simple as possible, just not simpler."

Queries and Sorts

Although the techniques outlined previously tend to be used more heavily when trying to develop and highlight more macro themes, you can often use the same techniques and data when you are trying to drill down for specific investment ideas. Sometimes as you move down to more micro levels in your analysis, you want to see specific details of debt instruments.

When you want to extract details from a database, you frequently use sorts and queries. Because of the multiple layers of descriptive data for corporate debt, the number of various iterations and sorting criteria that are utilized in a single project can become very high.

A good database allows you to run queries, for similar and varying characteristics. For example, suppose you are looking at opportunities to make intracapital swaps. You might want to ask your database to list all issuers of bonds that have both a Senior Secured and a Senior bond outstanding. To do this query, the database obviously has to be constructed correctly. It must have identifiers for each issuer and each ranking of bonds; it must also have up-to-date prices, yields, and spreads to offer the information you need. You might also want to limit the amount of information you get in your query. For example, you might want to only include U.S. dollar issuers and also limit it to bond issues over $200 million in size and might even want to see no bonds rated CCC or lower. There would need to be identifiers for all of these items within the database as well.

Sorting is basically what it sounds like. It is taking the data from your query and choosing how to sort it, trying to put it in an order to either answer a question or show a pattern. Building on the example in the prior paragraph, you might want to sort all issues that have a Senior Secured and a Senior bond outstanding and rank them by those with the largest spread between the two issues on a yield-to-worst basis all the way down to those that have the smallest spread between the two. This type of analysis might give you an idea which issuers have bonds outstanding that look most attractive for you to swap out of Senior to Senior Secured. The next layer would be to have a credit analyst explore the credit quality of the various issuers and the detailed terms of the bonds involved.

Queries and sorts can be very valuable to find specific recommendations and trade ideas. The key is to design them correctly and, of course, design the database so that it can be easily asked the queries. You will find queries are one of the most common analytical tools used, especially when doing analytics on a more micro basis.

Endnotes

1. When a regression is run, you can take the distance from a data point to the fitted line, the error, and square these (to account for positives and negatives) and then sum them. This is the standard error of the line (SE_{line}). The r^2 is 1-SE_{line}/SE_{avg}. The SE_{avg} takes the mean of all the data and calculates the error from the fitted line. One professor described this ratio as how well an intelligent prediction (the regression's SE_{line}) explains the data relationships as a percentage of how much a simple-minded prediction (the SE_{avg}) explains the data relationships.
2. Causation is an important concept to keep in mind regularly as you do analysis of relationships in the markets.

6

Data Mining

What Is Data Mining?

Some of the tools and terms used in data mining are from the school of statistics, such as regression, correlation, and standard deviation. Although some may define data mining as its own area of study, I view it as a branch of statistics.

Data mining uses the modern power of computers and algorithms and has received a fair amount of press. It is designed to take large amounts of disorganized and noisy data and try to discern patterns that can be predictive. Data mining can also be a good tool to find outliers from certain patterns.

Generally, when you lay out information in a graphical or sorted format, you are looking for specific relationships. When doing such a project, you might be designing the data to see if one sector of the market is over- or under-valued—or for some similar, specific factor. For example, a trading desk manager might want to see which types of bonds have regularly underperformed in his positions over a long time period. Perhaps he wants to see it by industry of the issuer as a category; thus, a series of queries could probably help to show this pattern. Typically in data mining, the system takes large amounts of data and tries to evolve patterns and relationships that might not be typically obvious. So whereas in traditional statistics you are specifically exploring a relationship, in data mining the system tries to show you patterns that you might not have initially been looking for or that were not obvious. Some of the techniques used involve theories such as cluster analysis, neighborhood, decision trees, and neural networks.

Many of the documented successes in data mining have been in marketing and sales. Data mining has taken large amounts of data about customers and found patterns. Every time you swipe your "special customer discount card" at a major supermarket or drugstore chain, they are gathering data that they can use to pump into their programs with complex algorithms and figure out how to sell you more goods at higher prices. One famous example often cited was a convenience store chain that discovered that on a certain day of the week, men would come in to shop and the two most common products that were purchased by the majority of them were diapers and beer, not two items usually placed near each other. So after processing the data, the store could change its product placement to maximize sales and adjust its pricing to make sure that beer was not discounted on those days of the week. Data mining took relatively random data and found a pattern that the analysis team might not have even been looking for. Certainly this relationship that the data mining team found was not something that would be intuitive, as a purchasing combination of diapers and beer is certainly not as intuitive as corn chips and salsa.

Data mining is not nearly as widely used in finance as it is in marketing and merchandising, although it does seem to be part of the next frontier in statistical analysis. Therefore, it is worth going over some of the basic techniques used in data mining, as it will likely eventually be more widely used in financial analysis and it might inspire you to think of new ways to use it.

Many different types of programs are used for data mining, one open source program that is a strong statistical package is called R, similar to the commercially available SAS or Statistical Package for the Social Sciences (SPSS). R has a graphic interface known as Rattle, which is used for many of these types of techniques.

Neighbors and Neighborhoods

One of the techniques that has been in use for some time in data mining is neighbors, or nearest neighbor, or neighborhoods. This is a technique that tries to predict something by taking records that have similar traits and assuming that if another item has these similar traits, then you could predict another item. For example, using a "neighborhood" as an example of this technique, suppose you look at a town and the majority of households that have an income above $100,000 and a lot size over 1/2 an acre and tend to own at least two cars. Supposing there is a block where all the lots were over 1/2 an acre and the households had incomes over $100,000, you could predict that the households would all own two cars. One area that this technique has been effective in has actually been in text recognition and retrieval, such as on news Web sites. This might be an area that could be used in the future for corporate debt analytics to help quickly analyze the text of financial releases, bond indentures, and loan agreements, or the changes in these documents.

Clustering

Clustering is another widely used technique. In its most basic form, clustering is somewhat intuitive. Think of it as if you had two shelves to put your dinnerware on. You might logically put all your bowls on one shelf and all your plates on the other shelf. However, just like in more complex clustering techniques, the programmer has to make some decisions regarding how clusters will be formed when items might not be as clearly defined—such as the decision you will have to make when you decide which shelf to put your large coffee cups on.

Tags are applied to each cluster. As more data is added, each tag gets a longer and longer list under it. Ideally, you would have many of these clusters and each tagged cluster would get larger and larger as you add more data. You would use algorithms to hopefully start to see what patterns of commonality would start to form underneath these tags. Clustering is often used not necessarily to see which data fits a pattern, but frequently to see which part of the data does *not* fit the pattern. For example, if one gas station in a chain of 100 stations is selling half the amount of gas, what other characteristics does it have that vary from the rest of the group? Perhaps all of the others are on a corner with a stoplight and this one is not, or maybe all of the others have a type of signage and this one does not. Like the old *Sesame Street* song, you are looking to see how, "...one of these things is not like the others, one of these things doesn't belong..." Perhaps in the future, this type of analysis could be used to identify a bond that has different covenant language relative to the other bonds in a complex capital structure, or it could be used to find one security within a larger universe that has a different trading pattern and to analyze why.

Clustering techniques can get significantly more complex with hierarchical clustering and multiple dimensions of clusters. But, hopefully, the previous examples give you an idea of the basic concepts behind clustering theories.

Decision Trees and Neural Networks

A decision tree is a tool that can be used in all kinds of problem solving and game theory situations. It is a very effective tool in credit analysis and also can be applied to larger data analysis in corporate debt. You take data from past events and apply the results of these past events to the probability weightings in a forward-looking decision tree.

Decision trees are based on a number of questions and answers, then taking each possible answer and applying the next level of questions for that answer level. One of the keys can be asking the right questions. For example:

- Is this company likely to get bought? Yes or no?
- Who is a likely buyer? Is it a bigger competitor, a vertical integrator, a smaller competitor, or a private equity firm?
- For each buyer, what is likely to happen to the bonds of this company? For example, do bonds get tendered for, do bonds get leveraged up and downgraded, or do bonds get deleveraged and get upgraded?
- What would be a likely trading level for these bonds under each scenario?

You can see from the preceding simple example how the question being asked is important, and the answers you choose are, too. Past data on the markets can be analyzed and can help with deriving the right questions, the right potential answers, and which probabilities should be applied to each of the possible answers.

During the last few decades, people working in data mining and statistics began using algorithms to help populate decision trees, sometimes basing them on vast amounts of data analysis. Theoretically, the algorithms should keep running until they end up with only one answer, or said another way, have no more questions.

Neural networks are somewhat like superdecision trees; of course, they are not actually neural, but are artificially designed to try to act like the neural system in a body. These series of algorithms try to "learn" from the data they are using by categorizing the data and designing decision trees from this data and multiple regressions on the data as well. Each series of questions can be considered a neural "node." In setting up the programs, the right questions need to be designed and then placed in the right nodes that need to be positioned correctly.. When using neural networks, the data often has to be prepared in slightly different formats than its original form. Typically, the data has to be in a relatively binary format so it can be used to answer the questions.

These types of systems have apparently tended to have more success in areas where you are looking to find an outlier or explain an outlier. They have also reportedly had success in finding a defining critical trait that results in certain outcomes.

Summary Comments on Data Mining

Thus far, data mining has seemed to get more use within the equity markets; I assume this is because the data is much more consistent, concise, and that it is more widely available. However, it is likely that over time some of these tools will creep increasingly into the typical corporate debt analysis toolbox. Market participants in the corporate debt market and in all investment asset classes are often looking for outliers and the cause of their unusual performance as well as defining characteristics that impact security prices. Data mining has had success in these types of projects. So, data mining techniques are likely to become increasingly common over time as a tool in corporate debt data analytics.

Closing Comments on Section II

Many statistical tools are commonly used in debt analytics, but there are many more that are not. Many participants barely use any statistical analysis at all. I think a greater use of statistical tools will help improve performance for participants, and lower risk.

When you are about to embark on an analytical project, try to think ahead as to what tools you might want to use—this should help you organize the data from the beginning and hopefully use the most advanced tools for your problem. Of course, to do all this most efficiently, you need to be sure to keep the goal of the project in mind, be aware of as many analytical tools at your disposal as possible, and communicate these goals with everyone on the team.

Section III
The Markets and the Players

By Robert S. Kricheff

This section of the book outlines the basic major sectors of the corporate debt markets. These include bonds and loans issued by corporations in the United States, Europe, and globally. There are differences between all of these markets, but the debts outstanding in these markets all require both debt analysis and corporate credit analysis.

The second part of this section outlines the various market participants in the corporate debt markets. We outline who these major players are and touch on how they utilize or should utilize data analytics in their work.

7

The Markets

Each segment of the corporate debt markets is defined by the types of debt, the ratings, or the currencies and country of domicile. There are similarities between all of the markets but also differences in risk and other characteristics.

Investment-Grade Corporate Bonds

Investment-grade corporate bonds (also known as high grades or IG) are generally the highest credit quality within the corporate debt markets according to the rating agencies. The price movements of these bonds tend to be more sensitive to any action taken by the rating agencies too.

- Investment-grade corporate bonds are bonds issued by companies that the two major rating agencies (Standard & Poor's and Moody's) rate as investment grade, defined as BBB- and Baa3 or better.
- If one major agency has an investment-grade rating on the debt and the other is below investment grade, it is said to be *split-rated*. Note that the corporate rating and the rating for an individual tranche of debt issued by that same corporation can be different.
- These bonds tend to have features similar to and trade more like noncorporate debt, such as government bonds and agencies,

and tend to trade more in unison with those markets. There is generally considered to be less credit risk than other areas of the corporate debt markets and more interest rate risk. Therefore, much of the analysis covered in this book tends to be focused on lower-rated tiers of the corporate debt markets, but it can certainly apply, and probably more regularly should be applied, to this higher-rated market.

High-Yield Corporate Bonds

High-yield corporate bonds (also known as leveraged bonds and junk bonds) tend to trade in a much wider range than investment-grade bonds and tend to be less sensitive to changes or actions taken by the rating agencies. High-yield bonds have been one of the faster growing areas of the corporate debt markets and, because of growth changes in this market, have been more dynamic than investment-grade corporate debt.

- High-yield corporate bonds are rated below investment grade by the major rating agencies. It also tends to include a reasonable portion of the market that is not rated by any agency (also known as NR).
- High-yield bonds tend to have somewhat more complex structures than those in investment grade, sometimes involving less-common interest rate structures and more commonly having call features. Because of the greater credit risk, these bonds typically have more detailed covenants within the bond indentures than investment-grade debt.
- Data analysis, as well as the related credit analysis, in this segment of the market tends to be more of a hybrid of traditional fixed-income data analysis and an equity-like component of analysis. There is more focus on credit quality and less on

interest rates in this sector because in many cycles and for many issues, changes in credit quality can vastly overshadow changes in general interest rates in impacting the performance of the debt.

Leveraged Loans

Leveraged loans are the loan equivalent of high-yield bonds. These loans are generally issued by banks, but not always purchased by banks, as there are mutual funds and other institutional investors that now buy bank loans. Like high-yield bonds, this market has been rapidly growing in size over the last few years.

- Leveraged loans are bank debt issued by below-investment-grade companies. Loans are not securities by definition, but leveraged loans look and act similar to high-yield bonds in many ways.
- Loans are typically the most senior-ranking debt in a capital structure. Loans tend to have several different features than are usually seen in bonds. Loans typically have floating-rate coupon structures and very limited call protection, but typically have more detailed covenants, often having maintenance tests. Loans also typically have shorter maturities than bonds.
- Loans are often issued in A and B tranches, with the A tranches having traditional bank structures such as amortization, whereas B tranches tend to resemble bonds more and often have fewer covenants and higher yields. There can also be a multitude of more tranches beside just an A and B.
- There is no listing of trades for loans and no public posting of them. Loans by their nature are all private and not publicly registered, and companies do not even have to make publicly available the detailed terms of the loan agreements. Trades in

loans typically have to be approved by the agent and/or issuing company. For all of these reasons, the bank loan market can be even significantly less liquid than the high-yield bond market.

Emerging Markets and International Bonds and Loans

Generally, the previous sections have referenced the markets for bonds and loans in U.S. dollars and the euro. These are some of the largest segments of the global corporate debt markets. Many companies domiciled in other countries have issued debt in these currencies as well. However, some investors and indexes exclude foreign domiciled companies from their universe, whereas some only exclude companies that are domiciled or do business in a country with a below-investment-grade rating on its national debt, which tend to be qualified as emerging market debt and often are categorized separately.

Many companies that are not in the Euro-zone and not in the United States issue in the currency of these two regions. However, another subsector of the international corporate debt markets is local currency issues, such as a Polish company issuing debt in Zloty as opposed to separately issuing debt in U.S. dollars or the euro. Investors and indexes also generally make distinctions between the currencies in which the debt is issued as well as the country of domicile.

All of these subsectors add many other ways to subdivide portions of the market and run data analysis when looking at the global bond markets.

Investing internationally, particularly for investors who are familiar with the U.S. standards, can present many issues. Different countries often allow for different accounting rules that can vary from international or U.S. generally accepted accounting principles, or GAAP, standards. Different countries might not require the same amount or level of financial reporting (for example, in the United States typically public companies are required to report quarterly and in Europe it is typically semiannually). These factors, as well as others, can lead to varying degrees of financial transparency, which raises risks for investors. Very importantly for debt investors, countries tend to have very different bankruptcy laws. Some are extremely fluid and unclear, and some are based on liquidation only and apply a very strict priority with no concept of reorganization, as in the United States. Obviously, debt being issued in various currencies also presents its own risks and another factor of concern. Emerging market debt also adds a heightened layer of political risk.

A notable difference when investing in emerging market corporate debt is that developments within a country's economy or within its political system can change very dramatically and rapidly and often override any credit performance factors. Political factors are particularly hard to analyze within the context of financial and investment analysis and can have a huge impact on investment returns—just ask anyone invested in Venezuela during the early part of the twenty-first century.

Despite all of these various differences and concerns in non-U.S. and non-Euro-zone investing, many of the tools and concepts outlined in this book are very valuable to use when investing or considering investing in corporate debt in these emerging markets. It should also be noted that despite bouts of high volatility, these international debt markets are growing considerably and are likely to be a larger part of the global corporate debt markets in the next decade.

Credit Default Swaps (CDSs)

CDS is used in both the investment-grade and high-yield bond markets. There is also CDS for loans referred to as LCDS. All work on the same basic principles. Since the 2008 financial crisis, there have been considerably more structure and regulation added to the CDS markets.

- CDS is effectively an insurance contract on a bond or a loan. The buyer of the CDS pays a regular fee to the seller of the CDS contract to guarantee that he or she will get par back for the principal of a bond or loan if the bond or loan defaults during the life of the CDS contract. In return, if a default occurs, the buyer of the CDS delivers the defaulted security to the counterparty on the CDS contract when he or she gets paid the cash.
- CDS is not available on every bond or loan; there has to be a willing buyer and seller to create the CDS contract.
- There are market participants who invest in CDS as a hedge to positions that they own on their trading desk, loan book, or portfolio and there are CDS investors who simply invest in the CDS as an investment unto itself.
- CDS written on a single credit is sometimes referred to as "single-name CDS" to distinguish it from CDS indexes. CDS indexes are made up of selected CDS single-name contracts. CDS trading and CDS indexes can have a meaningful impact on the trading movements of the assets that underlie the contracts; in other words, on the related bonds and loans. There can actually be more CDS issued on a company than it actually has debt outstanding. CDS data and CDS indexes can be analyzed as a tool to get a sense of how markets are performing as well.

8

The Participants

Introduction

It is worth understanding the different major types of players in the corporate debt markets and their different needs in using data analysis. I still believe many operators in all segments of the market have not advanced to using data analysis as much as they should, thus creating an opportunity for those who do and an opportunity for growth for those who supply and analyze data on these markets.

The various market participants have different types of data that they look for and their needs for data vary by what their role is and what parts of the market they operate within. For example, a trader at a sell-side broker-dealer has to be very transaction oriented and likely looks for significantly different data than a portfolio manager at an insurance company who tends to have a long-term, buy-and-hold style of investing. Or as a different example, given the limited data available on bank loan pricing, an analytical study on price movements might prove helpful, but the analysts would need to be sure all of the end users who the study was presented to understood its shortcomings.

It should be noted that the market for corporate debt instruments is by and large an institutional market and individuals are not typically involved in it. This is due to a number of reasons. First, *round lots* of trades are typically a minimum of $1,000,000 or an equivalent. The market is very illiquid from a trading perspective. Rules in the United States limit many of the debt issues to only be sold to "qualified buyers" and even some smaller institutions do not qualify. Individuals tend to be involved in these markets through mutual funds and ETFs or sometimes indirectly through their pension funds or insurance investments. Therefore, this chapter does not refer to the individual investor in the following sections.

The Issuers

The issuers of corporate debt are both private and publicly traded companies. Usually, the company has to be of a reasonable size—or at least the debt of the company has to be of a reasonable size—to make it into the general corporate debt markets (different people might have different breakpoints, but a minimum debt size of $100 million is not a bad cutoff line).

Most issuers of corporate debt only occasionally issue debt, perhaps in a three- to five-year cycle or even longer. Although their financial team will want to understand the markets in which they borrow, they are not likely to utilize detailed debt market analytics. However, there are companies that issue debt much more regularly, usually due to the more complex and multitiered structure of their capitalization or because they are very acquisitive by nature. Additionally, there are private equity firms that typically own stakes in a portfolio of companies and they usually use leverage to acquire these companies; therefore, these private equity firms are also, indirectly, very frequent issuers of debt.

These frequent issuers will want to analyze the data of new issuance in the markets; they will want to know what structures and in what industries new issues are being sold. Very importantly, they will want to know what yields investors are demanding for these new issue purchases, which could be used as a comparable for the type of issuance that they might be looking to do.

They will also want to gather and analyze data on overall supply and demand in the markets in which they want to issue debt and might want to compare the relative value of issuing debt in different markets, such as deciding whether to issue a leveraged loan versus a high-yield bond to fund an acquisition.

New issuance and related data and the ability to analyze it can be so important to some private equity firms that many have built their own capital markets teams that do this full time. Some also have set up new issue underwriting arms. Although it shows how important having a handle on new issue pricing is to these firms, it also might raise some serious conflicts—especially when many private equity companies now have asset management divisions with third-party money that invest in new issuance as well, while another arm is overseeing new issuance, underwriting, and otherwise influencing the paper that is being placed in the markets.

Investment Banks and Broker-Dealers

Investment banks generally act as advisors to the issuers, helping to guide them to what type of financing to have in place and what the costs are likely to be for the financing. Law firms and accountants are also intimately involved in this process. The investment banks get paid by the issuers.

Most major investment banks also have broker-dealer operations (or securities divisions). This part of the investment bank usually is involved in helping to distribute a new issue of debt to the buyers and then, using the firm's capital, helps to make markets in the debt instruments after they are issued. The capital markets group is the point person between the investment banking division and the sales force and is primarily responsible for handling the distribution and pricing of new financings for the investment bank's clients. The capital markets person needs to be able to balance various aspects of the market and various constituents when pricing a new deal. He or she needs a deal at a price that will trade well and has enough investor demand, but also gets the issuer the best pricing because both the buyers of the new issue and the issuing companies are clients of the investment bank.

To win the business of the issuer, the investment bank might have to put some capital at risk, fund a bridge loan (a temporary loan until the bonds and bank debt can be placed), or offer a backstop (guaranteeing a price for the issuer; otherwise, the bank will make up the difference). In addition, in bank loans the lead arranger of the loan is usually supposed to take a *hold,* meaning that the lead arranger keeps a part of the loan on his or her books. The lead arranger also frequently has to offer a revolving loan to the issuer, which is a relatively short-term line of credit that can be called on at the company's option.

The broker-dealer typically has a large trading desk that consists of research personnel, traders, and salespeople, all of whom work on new issues but also spend a considerable amount of time trading debt instruments in the secondary market (new issues are considered the *primary market* and other trading is considered the *secondary market*). In the secondary market, broker-dealers either use some of their own capital to trade securities or try to match buyers and sellers. Although post-2008 financial regulations started to put limits on the amount of capital that can be committed to owning securities long term by these firms, the firms still are always holding inventory as they try to provide liquidity to the markets, while still making money on the transactions. However, the amount that they can hold and how long they hold these positions has come under more and more regulation for broker-dealers who are part of the large investment banks. Additionally, regulations have raised the required capital reserves for all of these types of operations so the broker-dealers also have less absolute money to trade with and provide liquidity.

The investment banks that have broker-dealer operations have both the companies that issue the debt and the buyers of the debt as customers. This can create obvious conflicts. There is a maze of rules around conflicts and information flow as well as the needs to balance the business demands of all the constituents. It should be noted that there are broker-dealers that trade securities in the secondary market but do not participate in the primary markets, and there are investment banks that do little or no securities underwriting.

The investment banks and broker-dealers are often referred to as the *sell-side*. The sell-side is both a great user and generator of data and analytics. The sell-side has designed and manages all of the most followed indexes in the debt markets. It also often has strategists who run macro strategies and try to develop investment ideas for their trading desks and their trading desks' clients.

The investment bankers and the capital markets personnel are extremely interested in data that captures how other new issues have been priced and traded as well as what structural features they have. They use this information to advise the clients issuing the debt where their bonds or loans might price and how different features will impact that pricing. This is similar to the data that frequent issuers would want to obtain and analyze. The information would pertain to how comparable transactions are trading in the secondary markets. It is also important for them to monitor general supply and demand in the markets. It can also make a difference to them the type of money that might be coming into the corporate debt markets; for example, new funds from collateralized loan obligations, insurance companies, or hedge funds all favor different types of issuers. There is a whole school of tracking and analyzing the supply-and-demand data, which is generally referred to as market technicals.

The credit analysts, traders, and salesmen at broker-dealers use data analysis for a number of things. The technicals provide a guide as to the near-term market trends and help give them an idea if they should increase or decrease their capital committed to the markets; most trading desks also run some types of hedging and these strategies are impacted by the technical analysis, too. They also use data for more microanalysis to try to analyze relative value among various sectors of the market and then down to specific bonds or loans; they do this to know what they might be more or less willing to own in their inventory but also to advise their trading and sales clients.

At the manager level, and hopefully below, risk analysis should be constantly being run. The risk analysis needs to analyze the size and age of trading desk positions and weigh these versus other possible capital commitments within the firm and the risks and trading liquidity of these positions. Keep in mind that trading desks have a cost for the capital that they use and this needs to be factored into the risk metrics. Additionally, forward-looking sell-side firms analyze data about which traders and desks are using how much capital and their return. These firms also periodically review their business to see the level of trading with various clients, or lack thereof, as well as the profitability of those trades to help manage their business.

As in all of these categories, the level to which participants use data and analysis to help improve their business varies greatly from firm to firm and also by individuals.

A Note about the Sell-Side

One major characteristic of a sell-side investment bank, from the corporate finance department through to the salespeople, is that it is transaction oriented, whether it is the placement of a new issue or the crossing of secondary trade between two clients. Because the firms are so transaction oriented, resources dedicated to some of the analytical tools can be very limited, if they cannot immediately be related to transaction revenue or tied into new regulatory requirements. The other outstanding characteristic of the sell-side versus other major participants is that its capital tends to be much more short term in nature than its client base. Liquidity and turnover of its capital are key risk metrics.

Money Managers and Institutional Investors

The trading and sales clients of the broker-dealers are money managers, also known as asset managers or the *buy-side*. The makeup of the buy-side varies much more greatly than the landscape of the sell-side. Asset managers vary in the type of money they manage, their management styles, their fee structures, and their own corporate structures.

Some examples of the types of money they might manage are as follows:

- Pension assets
- Mutual funds, which can have strategies that run the gamut
- Insurance assets
- Family wealth funds
- Sovereign wealth funds
- Structured products such as collateralized debt obligations

These assets can be managed in different styles from extremely conservative, perhaps requiring only investment-grade ratings and not allowing any net losses to be taken, to strikingly liberal or aggressive rules, such as requiring no BB-rated bonds and allowing equities and defaulted securities within the investment basket.

The money can be managed internally, such as through an internal investment team at an insurance company, or can be farmed out to a stand-alone investment manager. Often when the money is given to a stand-alone investment manager, it is being given to fulfill a specific diversification desire. For example, a family wealth fund might choose to diversify its assets by putting 40% of its investments in stocks and 60% in fixed income. It might then want to diversify that fixed-income basket among 10% investment-grade, 60% core high-yield bonds, and 30% in an aggressive hedge fund strategy. These three strategies may be spread between three different managers or could all be given to one manager to allocate between the three different strategies. However, the family wealth fund decided on the allocations; it does not want the money manager who is handling core high-yield bonds deciding that investment-grade bonds are a better buy right now and investing in that market, nor does the fund want large positions in cash sitting at the managers it allocated investments to. So the asset manager's job is to invest those assets within the class that they were allocated for.

Asset managers should be very large users of data analysis. They should use systems to monitor a number of administrative items. Theses administrative items should analyze the rules of the accounts they are managing and that the investments and trading are following those rules. Other administrative analysis may cover how much turnover there is in an account or the amount of trading volume that is done with various broker-dealers to make sure there are not unreasonable biases.

As far as the data analysis for investments, the possibilities seem endless, but there should be a distinct discipline of regular, vigilant review of certain key data points and trends. The key items may vary from client to client and asset manager to asset manager, but they should be reviewed. Some of the typical items that should be regularly analyzed by asset managers include the following:

- Development of macro themes
- Sorts and queries to find investments that fit the macro themes

- Performance versus a benchmark
- Weightings versus a benchmark
- Performance attribution of the portfolio
- Key characteristics of the portfolio
- Characteristic shifts over time in the portfolio
- Dispersion within and among portfolios
- Volatility of performance within a portfolio
- Betas and alphas of holdings in a portfolio
- Sensitivity to macroeconomic factors in a portfolio

These are just some of the types of data analysis that might be undertaken by buy-side managers. The buy-side managers might also care about the detailed technical and new issue analysis that the sell-side focuses on. Understanding the technicals of the market can be the key for money managers in timing buys and sells when possible. There are cycles when the technicals of the marketplace overwhelm all other considerations.

It is important to keep in mind that many of the administrative systems and analysis are required in the business, but many of the analytical tools of data analysis are not always invested in by asset managers, even though it will make their performance better in the long run.

Buy-side and sell-side firms typically use a combination of semi-customizable turnkey programs as well as in-house developed programs. Most firms in the bond business also use the Bloomberg system, which not only has bond calculators and news and chat functions, but also offers trading functionality and portfolio management tools. It is embedded in virtually every bond operation. However, there are many other systems that people use for portfolio management and trading inputs as well.

Asset Allocators and Consultants

Another important part of the industry is asset consultants. These professionals work very closely with pension allocators and family wealth offices and similar asset holders. The consultants analyze and work with their clients to choose which fund managers they want to allocate money to. They help determine which asset class they want to invest in, what weightings they want to pursue, and what type and style of investment professional they want handling the money in that asset class. Sometimes the clients (that is, the pension fund or family wealth office) manage some of the assets internally and some externally, or they might farm them all out externally as it can be more cost effective and give them more flexibility over time.

The consultants use numerous analytical tools to decide on diversification strategies as well as to weigh the risks and returns of the potential asset managers to allocate money to. On an ongoing basis, they then analyze the performance of these managers.

Systems Managers and Programmers

The people who design these programs and systems with which to analyze the bond data have various seats in buy- and sell-side firms. Sometimes they sit in the information technology (IT) department, sometimes they are part of a research or strategy group, sometimes they are programmers and analysts embedded on a trading desk or in a risk management department, or sometimes they might be a part of a portfolio analysis team. All tend to have a formal, or informal, computer programming and/or systems skill set. Ideally, they need to be collaborative with the end user and have an understanding of the basic goals and tools that are being used by the end user. Ideally, they have good visual presentation skills so that once they understand the end goal of the project and what they are driving at, they can lay out data in a format that addresses the main question but also supplies the other information that the end user may want available when analyzing the data.

Closing Comments on Section III

The investment in data and analytics in the corporate debt markets has been large. Some of the initial investments have just been put in place to track the complex and ever-changing data on the terms of each security cross-referenced with some of the details of each issuer. However, more and more, the analytics that are required involve detailed data that combines traditional bond and equity information and often ties in macroeconomic scenarios as well.

The different participants within this market have differing goals and uses for this data and are making various levels of commitment to data analytics. I do believe the demand for people with a programming and systems background who understand the types of analytics that are being used will increase. This book I hope can be particularly helpful for these professionals in their interactions with end users and investment professionals.

Section IV
Indexes

By Jonathan S. Blau

Indexes are one of the main tools used in analytics. They help to give a snapshot of what a market and market subsectors look like. They are often used for performance and weighting measures.

As a significant amount of analytics is done using indexes, it is important to understand some of their strengths and weaknesses. It is also important to understand how they are constructed so you can better understand the biases of the various indexes that are available.

Unlike many equity market indexes, there is no single, dominant index that is used in the corporate debt markets. Additionally, all of the major indexes are run by sell-side investment banks.

9

Index Basics

Why Do Indexes Matter?

Virtually all securities markets use indexes to measure performance. The best-known use of an index is as a benchmark to measure the performance of a portfolio of securities from the same market. The index measures characteristics such as return on investment, *total return,* expected future rate of return, *yield,* or expected future return above the risk-free rate, *spread.* Each of the measures of an index is an average across the component securities of the index.

Indexes are also used to answer questions about both the behavior of the market and the portfolio managers who manage funds of the market's securities. Some of these questions are as follows:

- How does the current market compare with historical periods? The index can help us predict the future behavior of the market by comparing the current characteristics with what happened after a previous period with similar characteristics. How is the current period similar to history? How is it different? How might these similarities and differences change the outcome? Exhibit 9.1 illustrates the changing characteristics of the high-yield market from 1996 to 2012 by comparing the characteristics of a high-yield bond index over that period.

The relationships of these characteristics over time can be related to events occurring outside of the market. Corporate debt markets are sensitive to interest rates, liquidity in the financial system, and the strength of the economy.

Exhibit 9.1 Comparison of characteristics for a generic high-yield bond index

	(At Year-End Except for Total Return)				
	1996	**2000**	**2004**	**2008**	**2012**
Face value (billions)	$315	$595	$746	$812	$1,102
Market value (billions)	$316	$417	$755	$499	$1,157
Total return	13.2%	–5.8%	12.4%	–25.8%	14.4%
Yield to worst call	9.5%	15.2%	6.7%	19.2%	6.4%
Spread to worst call (bps)	335	971	328	1,771	562
Current yield	10.0%	13.5%	7.9%	13.1%	7.6%
Yield to maturity	9.8%	15.1%	7.4%	19.8%	7.3%
Modified duration (in years)	4.1	4.4	3.9	3.8	3.5
Average rating	B+	B+	B	B	B+

- Is the market driven by systematic or idiosyncratic risk? *Systematic risk* is behavior seen in groups, or *sectors,* of securities, such as all of the securities in the same industry or with the same credit rating. *Idiosyncratic risk* is behavior driven by one or a few specific securities, independent of which sectors those securities are classified in, such as a large issuer in a sector missing its expected operating results or getting upgraded by the credit agencies. Exhibit 9.2 shows some typical sectors used in corporate bond indexes.

  The index should provide enough information so you can determine which sectors or securities are driving the index results, measures such as return, yield, or spread. This includes the top-level results for the index overall, the results broken out by sector, and the list of the component securities of the index with their individual results and the sectors into which each security has been classified.

Exhibit 9.2 Typical sectors used in corporate bond indexes

Sector	Comments
Industry	Macroeconomic driver
Credit rating	Typically a blend of two or more agencies' ratings
Size of issue	Indicates liquidity of issue; typical categories: under $300 million, $300 to $500 million, $500 million to $1 billion, etc.
Duration/years to maturity	Term structure; typical categories: less than 1 year, 1 to 4 years, 4 to 7 years, etc.
Price	Used in below investment grade to indicate risk; typical categories: under 50, 50 to 75, 75 to 90, 90 to 100, above 100
Seniority	Security ranking: secured, 2nd lien, unsecured, subordinated
Coupon payment type	Cash-pay, stepped coupon, zero coupon, pay-in-kind, defaulted
Use of proceeds	General corporate purposes, refinancing, acquisition, or buyout
Country/region	For indexes across multiple jurisdictions
Currency	For indexes with issues denominated in multiple currencies

- How does a portfolio selected by a fund manager differ from the index, and what does that tell us about the manager's skill? When managers use an index as a benchmark, they can control their portfolio's exposure to systematic risk by balancing the relative size of the sectors in the portfolio as compared with the index. For example, they may underweight industrials or overweight BB-rated bonds, as compared with the weights in the index.

  Controlling the systematic risk of a portfolio, or *sector selection,* requires a view on how sectors will perform. To outperform based on sector selection, a manager needs to overweight the sectors that he or she believes will outperform the overall index and, likewise, to underweight the sectors that will underperform the index. This is an important skill for a manager, but it is the relatively easier task. Once the manager has a view on sector performance, adjusting weightings is not difficult because there are often a large number of securities to choose from to increase or decrease the exposure to any sector.

The most-valued skill of a manager is to outperform the benchmark based on idiosyncratic risk, or *security selection:* the manager's ability to pick specific securities that outperform without changing the systematic risk of the portfolio. For example, the manager might want to underweight industrials while still owning some securities in that sector. Picking those specific industrial debt issues changes the idiosyncratic risk without affecting the systematic risk. One factor that also has to be considered is what level of risk or volatility the manager is willing to take to reach these returns.

- How does one market compare with another? Indexes of different markets can be used to compare the relative performance of the markets. The most common characteristic to compare is the total return over the same time period for two different indexes. This is because the calculation of total return is the same in any market: It is the total change in value over a time period.

  Comparing only the total return of indexes from two markets leaves out an important consideration: How are the returns comparable while taking into account the relative risks of the markets? Indexes can help answer this question as well. One simple measure of risk is to calculate the volatility of total returns. For example, this can be calculated as the standard deviation of monthly total returns over a period. By dividing the total return of the period by its volatility, you can calculate the return per unit of risk, as seen in Exhibit 9.3. There are a number of more sophisticated techniques to measure risk-adjusted return, but they all use this basic concept.

Comparisons can also be made using measures such as yield and spread, but the methodology for computing these measures often differs between markets, making comparisons much more difficult. It is sometimes possible to adjust these measures to make the comparison fairer, but this is often a very complex exercise.

Exhibit 9.3 Return and volatility: 2003 through 2012

Market	Total Return	Volatility	Return/Risk
U.S. treasuries 5–7 years	5.4%	5.2%	1.0
Investment-grade bonds	6.5%	6.7%	1.0
Large-cap equities	7.3%	15.5%	0.5
Mid/small-cap equities	9.7%	21.5%	0.5
Emerging markets	12.0%	10.0%	1.2
High-yield bonds	10.2%	10.6%	1.0
Leveraged loans	5.5%	7.9%	0.7

Notes: Total return is the annualized monthly returns. Volatility is the annualized standard deviation of monthly returns.

Calculation Methodology

The index results—such as the total return, yield, and spread—are averages of these measures of the individual securities—the *constituents*—of the index. Most of the averages are weighted by capitalization, or market value, of the constituents. This is the same methodology used in calculating these measures in a portfolio of securities, which permits you to use the index as a benchmark because the comparisons are on the same terms. The *market weight* is the face value (par value) of the security multiplied by its price, then divided by the total market value of the index.

Why do we use market weight? Suppose you have only two investments, and one has ten times the value of the other. That larger investment will have ten times the impact on the overall outcome as the smaller investment. Market weight captures this methodology.

For measures that depend on the price of a security on a date, such as the yield or spread, you compute a weighted average using the market weight of each security on that date, which is called the *market weighted average.* For example, when reporting the yield of the index for a month, you calculate the market weighted average of the yields of each security on the last day of the month.

Total return is a measure that depends on two dates, the start and the end of a period. The total return is the change in value over the period: the change in price from the start to the end plus the coupon interest paid or accrued during the period. Because corporate bonds are usually transacted with the accrued interest, the formula for total return is:

$$\frac{(\text{Ending Price} + \text{Accrued Interest}) - (\text{Starting Price} + \text{Accrued Interest}) + \text{Coupon Paid (If Any)}}{\text{Starting Price} + \text{Accrued Interest}}$$

You compute the average total return of the index weighted by the market weight of each security at the *start* of the period, not the end of the period as you do for yield or spread.

To put this in context, suppose you are calculating the total return of the index for a month. Imagine that you buy all of the securities in the index at the start of the month and then sell them at the end of the month. (We always use prices on the same side of the market, usually the bid price, so as not to include transaction costs in the index.) The total return is the change in value after this transaction, including the accrued and paid coupon collected. The cost of each security is its market value at the beginning of the month. The average outcome you receive is the average of each security's total return weighted by this starting cost. In other words, the total return is weighted by the market weight at the start of the month.

For some measures, you may calculate an average based on the *par weight*, which is the face value of a security divided by the overall face value of the index. Par weight is used for calculating averages where the contribution of each security is independent of its price. One such measure is the average price of the index: The par-weighted average gives you the average price you paid to buy all the securities in the index.

Another use of par weight is to compare the weight of highly discounted, or *distressed*, securities with the entire index. Distressed securities can be defined in a number of ways, such as securities below a maximum price, above a minimum yield or spread, or rated at or below a distressed rating level, such as CC. Managers who specialize in portfolios of distressed securities might want to illustrate the size of the investment opportunity to their investors by using the par weight of those securities in the overall index, even though the calculation of return will be by market weight.

A third example is computing the default rate, important in below-investment-grade markets. The *default rate* is the par weight of the defaulted securities. You use par weight because there is a separate measure, the default loss rate, which uses the price of the defaulted securities combined with the default rate.

Index results are computed at a particular *frequency*. Many fixed-income indexes are computed and published daily, like equity indexes. But in many markets, especially for corporate debt, prices for many, if not all, securities are available on a less-frequent basis. For example, prices might be made available only once a month for many of the securities in an illiquid market, like bonds traded in the structured finance market. An index of those bonds can't be computed more frequently than monthly.

Because the constituents of an index can be divided into groups based on characteristics, like industry or rating, you compute the same results for these groups, or *sectors*, as you do for the overall index. Examining the results of an index by sector is a method to determine the systematic risks driving the index. Which industries have a higher yield than the index overall? Did large issues underperform small issues? How did the spread of BB securities change as compared with CCC securities? Answering these kinds of questions helps explain the market behavior.

Sectors are also used to compare the systematic risk of the index with a managed portfolio of securities. A technique called *attribution analysis* compares the relative performance and the relative weights of the sectors of the index and the portfolio. This technique isolates the systematic risk of the portfolio. Exhibit 9.4 is an example of attribution analysis.

Using this technique, the "excess" performance, not explained by systematic risk, is the idiosyncratic risk of the portfolio.

In financial literature, systematic risk is referred to as a portfolio's *beta* (β) and idiosyncratic risk as its *alpha* (α). The equation used to express the return of a portfolio is often stated in a simplified, linear form as

$$\text{Return of portfolio} = (\beta \times \text{Return of index}) + \alpha$$

By this definition, when an index is used as a benchmark, its beta is 1 and its alpha is 0.

Exhibit 9.4 Example of attribution analysis: generic high-yield index versus managed portfolio

	Index		Portfolio		Index vs. Portfolio	
Industry	**Weight**	**Return**	**Weight**	**Return**	**Weight Difference**	**Return Difference**
Consumer	18.7%	1.1%	16.2%	1.0%	2.5%	0.1%
Energy	19.2%	0.8%	14.9%	0.6%	4.3%	0.2%
Industrials	25.8%	1.0%	31.6%	1.2%	–5.8%	–0.2%
Media/telecom	15.5%	0.9%	13.7%	0.8%	1.8%	0.1%
Services	20.8%	1.2%	23.6%	1.4%	–2.8%	–0.2%
Rating						
BB	31.5%	0.8%	54.9%	1.4%	–23.4%	–0.6%
B	55.8%	1.1%	42.6%	0.8%	13.2%	0.3%
CCC	12.7%	1.3%	2.5%	0.3%	10.2%	1.0%
Seniority						
Secured	24.1%	1.1%	16.7%	0.8%	7.4%	0.3%
Unsecured	71.1%	1.2%	76.6%	1.3%	–5.5%	–0.1%
Subordinated	4.8%	0.6%	6.7%	0.8%	–1.9%	–0.2%
Size of Issue						
Under $300 million	11.6%	1.1%	3.2%	0.3%	8.4%	0.8%
$300 to $500 million	22.0%	1.2%	22.5%	1.2%	–0.5%	0.0%
Over $500 million	66.4%	1.0%	74.3%	1.1%	–7.9%	–0.1%

10

Index Construction

Introduction

How do we construct an index? An index is a portfolio of securities selected from the market by criteria that are designed to measure particular attributes, or *risk drivers.* The securities can be selected from all of the securities in the market, or from a subset that reflects a set of investment preferences. Some examples are liquid securities, securities with a particular credit quality, or securities with parameters like a minimum dollar price or a maximum yield or spread.

The first step is to determine the objective of the index. Is the index intended to be used by market participants who want to understand and measure the market overall? Or, is the intent to have narrower criteria to measure only a subset of the market? The risk drivers to be measured and the objective provide the context for the selection criteria. Exhibit 10.1 shows some common objectives for corporate bond indexes.

As part of the index objective, a clear definition of the market from which the index selects securities is necessary. Markets are usually defined by the buyers and sellers of a set of securities with a discrete set of characteristics. The securities can be traded on an exchange or over-the-counter. Some markets are defined geographically and can contain securities denominated in one or several currencies. Exhibit 10.2 contains definitions for some common fixed-income markets.

Exhibit 10.1 Examples of objectives for corporate bond indexes

Objective	Sampling Criteria
Entire market	All characteristics are sampled, such as industry, rating, issue size, seniority, etc. For markets with a large dispersion within sectors, indicating a high degree of idiosyncratic risk, every issuer is sampled.
Liquid securities	Issues larger than a specific size are sampled, such as larger than the average issue face value. Within this subset, all characteristics are sampled, such as industry, rating, etc. Selection is often determined subjectively by a committee of traders or other specialists.
Credit quality	Issues are limited to one or a few rating categories, such as all securities rated above B. As an alternative in below-investment-grade markets, limit to a price band, such as all issues in the top or bottom quartile of the price distribution.
Term structure	Issues are selected within a maturity or duration band, such as issues maturing in less than three years.
Seniority	Issues with a specific security ranking are selected, such as secured issues.

Exhibit 10.2 Some fixed-income markets

Market	Definition
Treasury bonds	U.S. Treasury bonds
Mortgage-backed securities	Agency and non-agency mortgage-backed pass-through securities, including both fixed- and adjustable-rate mortgages
Municipal bonds	Long-term, tax-exempt bonds, including state and local general obligation and revenue bonds
Investment-grade bonds	Corporate bonds rated BBB- and higher (rating determined as a blend among the agencies)
High-yield bonds	Corporate bonds rated BB+ and lower (rating determined as a blend among the agencies)
Leveraged loans	Tradable bank loans rated BB+ and lower; if unrated, loan facilities with margins above a minimum, such as 150bps (basis points)

The securities must have transparent characteristics that any market participant can determine. For corporate bonds, this includes the coupon or margin, maturity date, and other terms that are required to calculate the index measures, such as return, yield, and spread. The issuer can be classified by industry and by other characteristics, such as credit ratings and security ranking (seniority) in the capital structure of the issuer. As mentioned earlier, we call these classifications *sectors.* (Some market participants use *sector* to mean only the industry classification, but we are adopting a broader definition to encompass all of the classifications of the security.)

The risk drivers of the index are one or more of the sectors that the index measures. In corporate bonds, industry, rating, size of issue, and duration are usually identified as risk drivers, but this varies by market.

The market must have an established method to provide observable valuations, or prices, of its securities to the market participants. Price discovery in any market ultimately requires capital: The price of a security is really only known when a transaction between a buyer and seller occurs. On exchanges, specific rules exist for how prices are discovered and how that information is disseminated to market participants. However, most corporate debt instruments do not trade on a defined market.

Because most corporate bonds and all loans are traded in over-the-counter markets, prices in these markets are collected and disseminated by an established market trading system or from commercial data providers, which collect and aggregate prices from all of the dealers in the market. In these markets, bid and offer prices are shown by dealers to customers in anticipation of a transaction. In less-liquid markets with relatively few transactions, these prices, often called dealer *marks,* are often disseminated and used by market participants, including indexes, in lieu of actual transaction prices.

Securities priced using dealer marks may be updated less frequently than the prices of the more-liquid securities because not every mark may be updated every day. Eventually, transactions in the less-liquid securities occur and dealers update the prices. Indexes that contain these securities are no more accurate than the underlying pricing that is available. In other words, even when the index frequency is daily, if a significant weight of the index is priced using dealer marks, the daily results will reflect some valuations that were determined before today.

Index Construction: Selection Criteria

The selection criteria of an index can be *rules-based*, where the constituents are selected by rules that are established at the outset and infrequently, if ever, changed. In this method, the rules specify the characteristics of the securities to be added and removed, the frequency of updates, and the dates that changes are implemented. The rules are specific and clear so that the composition of the index does not require subjective judgment to implement. Rules-based indexes can be very broad samples of the market, or they can be designed to sample narrower criteria of interest to a subset of market participants. Exhibit 10.3 shows some typical selection rules for indexes of corporate bonds and loans.

Exhibit 10.3 Example of selection rules for corporate bond indexes

Market	Selection Rules
Investment grade	U.S.-domiciled issuers, USD-denominated issues include the most liquid bond at each tenor for each issuer rated BBB- and higher, based on a blend of the agencies' ratings at least $250 million in outstanding size at least one year until maturity bullets only (no amortizing bonds) fixed-rate and step-up coupons (no floating-rate issues) Securities Exchange Commission (SEC) registered and Rule 144A issues issues added and removed at start of month

Market	Selection Rules
High yield	Issuer domiciled in developed country, USD-denominated issues rated BB+ and lower, based on a blend of the agencies' ratings include at least one bond of each issuer; if more than two, include the two largest by size at least $100 million in outstanding size at least one year until maturity fixed-rate, step-up, pay-in-kind and zero coupons (no floating-rate or convertible bonds) SEC registered and Rule 144A issues defaults remain in index until emergence from bankruptcy new issues added from first trade date issues downgraded from investment grade added at beginning of next month after downgrade issues removed upon call, tender, or emergence from bankruptcy
Leveraged loans	Issuer domiciled in developed country, USD-denominated loans rated BB+ and lower, based on a blend of the agencies' ratings include all outstanding fully funded facilities of each issuer at least $50 million in outstanding size at least one year until maturity floating-rate coupons, with and without LIBOR floors defaults remain in index until emergence from bankruptcy new issues added from first trade date issues removed upon paydown or emergence from bankruptcy

An example of a broad rules-based index is to include one or more issues of every issuer in the market. Every issue over a certain size, for example $100 million face value, can be selected. Very small issues may be so illiquid that they cannot be priced at the same frequency as other issues. Or, the selection can be narrowed by picking the most liquid securities of each issuer, or by picking at least one issue at each rating or seniority level of each issuer. A broad-based index is designed to provide insight into broad market behavior, and it can serve as a benchmark to portfolios that can purchase any market security and may hold the security for a considerable period.

An example of a narrow rules-based index is one that includes only issues of a specific rating category, such as only the BB-rated securities of the market. Such an index would be useful to the investors who intend to invest in only those securities.

Another method is to select the constituents of the index by *subjective* criteria, as determined by a committee of market participants. In selecting the constituents, the committee may use some rules that define the objective of the index. But a portion of the criteria is based on a subjective judgment or factors that are not easily determined by a fixed set of rules. The participants suggest securities to add and remove based on their subjective judgment. The committee establishes rules to make its determinations, such as voting procedures and thresholds. The committee also establishes the rules to determine the frequency of updates and the dates when the changes are implemented. Because the criteria are specific and subjective, these indexes often measure a subset of the overall market.

For example, a committee can specify an index containing securities that are most actively traded. This index is useful to the traders in the market to compare the index results with the P&L of their trading books. It can also serve as a benchmark for portfolios that are managed with an active trading strategy, where most or all of the securities in the portfolios are among the most actively traded.

Whatever the method, the selection criteria can be designed to cater to a particular fund investment style so that the index can be used as a benchmark for funds of that style. For example, an index that only contains issues of short maturities may be designed to serve as a benchmark for funds with a similar short maturity investment criteria.

Index Construction: Requirements

Whatever selection process is used, the index should reflect several broad requirements.

The index should reflect a neutral, unbiased investment. This means that the set of securities selected should capture all of the systematic and idiosyncratic risks of the securities within the criteria. In practice, every index has biases inherent in its selection criteria. Neutral means that the criteria are evenly applied across all of the securities in the market so that there are no implicit criteria biasing the results. As an example, an index for the most actively traded securities should be determined by a majority of market dealers, not be limited to the securities traded by only one or two market participants. Exhibit 10.4 compares two high-yield market indexes, one that contains every observable issue in the market, labeled "All Issues Index," and the other that includes only the most liquid, frequently traded issues, labeled "Liquid Index."

Exhibit 10.4 Comparing two high-yield market indexes

June 2013

	Market Weights		*Difference*
Industry	**All Issues Index**	**Liquid Index**	**All vs. Liquid**
Consumer	17%	11%	6%
Energy	17%	19%	-2%
Industrials	26%	25%	1%
Media/telecom	18%	22%	-4%
Services	22%	23%	-1%
Rating	**All Issues Index**	**Liquid Index**	**All vs. Liquid**
BB	38%	37%	1%
B	51%	52%	-1%
CCC	11%	11%	0%
Seniority	**All Issues Index**	**Liquid Index**	**All vs. Liquid**
Secured	21%	27%	-6%
Unsecured	74%	68%	6%
Subordinated	5%	5%	0%
Duration	**All Issues Index**	**Liquid Index**	**All vs. Liquid**
Under 4 years	58%	48%	10%
4 to 7 years	32%	43%	-11%
7 to 10 years	7%	8%	-1%
Over 10 years	3%	1%	2%

The all issues index includes every observable issue in the high-yield market. The liquid index includes only the most liquid, frequently traded issues.

Each security included in the index must be classified into various characteristics, corresponding to the risk drivers that the index is trying to measure. For example, corporate bonds can be classified into the sectors industry, rating, and seniority (to name a few). Each security must be assigned into every sector used by the index so that it can be included in the calculations for each sector. Note that to use the index for attribution analysis on a portfolio, the securities in the portfolio must also be classified into the same sectors because attribution analysis compares the index with the portfolio using these same sectors. Exhibit 10.5 shows typical sectors used in corporate bond indexes.

Exhibit 10.5 Typical sectors in corporate bond indexes

Industry	Rating	Coupon Payment Type
Airlines	AAA	Cash-pay
Automobiles	AA	Zero coupon
Chemicals	A	Pay-in-kind
Consumer products	BBB	
Defense	BB	**Seniority**
Energy	B	Secured (1st lien)
Financial	CCC	Secured (2nd lien)
Food	CC	Unsecured
Gaming/lodging	Defaulted	Subordinated
Health care		
Industrials	**Region**	**Tenor**
Media	United States	Under 4 years
Packaging	Canada	4 to 7 years
Paper	Western Europe	7 to 10 years
Real estate	Other developed countries	10 to 30 years
Retail	Asia	Over 30 years
Services	Central/Eastern Europe	

Industry	Region	Use of Proceeds
Steel	Latin America	General corporate
Technology	Other developing countries	Refinance
Telecom		Acquisition
Transportation		Buy-out
Utilities		

Therefore, there must be sufficient information available to classify each security into each of the classifications in order for that security to be eligible for inclusion. This might seem like an obvious requirement, but some markets have limited transparency for some of the securities in the market. For example, in the leveraged loan market, some features of new issues might only be partially disclosed in public sources, such as the face value ("borrowed amount") or the coupon (for loans, this is the margin paid above LIBOR). Such issues are not eligible for inclusion in a leveraged loan index because insufficient information is available.

On occasion, a new characteristic may be identified and a sector for that driver is added to the index. Another example from the leveraged loan market is the feature of the LIBOR floor. Since 2008, many new loans have included this feature, which helps determine the coupon. If the current rate of LIBOR is less than the LIBOR floor, the floor is used instead, and the coupon is calculated by adding the margin to the floor. Indexes for the loan market needed to add this characteristic to each of the loans in the index with a floor. The classifications "has a floor" and "doesn't have a floor" became new sectors for calculating index results.

Ideally, added classifications are computed not only going forward, but also historically from the index inception, permitting comparisons of the new characteristic with the past. This creates an additional challenge: reclassifying the new characteristic for all of the securities that were in the index historically.

The securities eligible for inclusion must have observable prices. This might also seem like an obvious requirement, but the availability of pricing is very often the limiting factor in creating and calculating indexes. Many fixed-income markets contain securities that are priced at least once a day, and indexes based on those markets are computed daily.

However, some markets contain illiquid securities and in others, many (or most) of the securities are rarely traded. It is possible to create indexes for these markets, but the availability of prices must be taken into account when determining the frequency of the index. For example, if most of the securities in the market are priced only once a month, an index of those securities cannot be expected to provide results more than monthly. If the market is generally liquid with daily prices, but a few securities might not be consistently priced every day, even with trader marks, those securities might fall out of the index because of the lack of pricing, even if the security is selected for the index based on its criteria.

For corporate bond indexes, even ones that produce daily results, the results at the end of the month are very often the most important to market participants. The monthly results are the most useful when trying to use the index to understand the market's behavior or to compare one market with another because a month is a common yardstick in measuring activities in the capital markets. Also, investors using a fixed-income index as a benchmark often look at the monthly results when measuring against a portfolio.

The index sample must be broad enough so that there is statistical significance both overall and by sector. If a sector has only a few bonds or a very small weight in the index, it ends up measuring only the idiosyncratic risk of those bonds, not a systematic risk of the market. When examining the index results by sector, it is important to look at the weights of each sector in addition to results like total return, yield, or spread. If a sector has a very small market weight, for example only a fraction of a percentage point of the market weight, the results should be understood to be idiosyncratic.

For example, the high-yield bond market has just a few securities that mature in longer than 20 years, less than 1% of the market by market value. These are mostly "fallen angels," issues that were originally investment-grade-rated and were downgraded to high yield. In comparison, the investment-grade market has many securities with maturities longer than 20 years. A common sector for both high-yield and investment-grade indexes groups securities by the number of years from now to the maturity date. If one of those categories is bonds that mature in more than 20 years, the sector will contain many bonds in the investment-grade index, but very few in the high-yield index. Drawing conclusions based on this sector's results in the high-yield index can be misleading because it captures only a small number of issues with a small weight in the index. In the high-yield index, the sector is measuring an idiosyncratic risk. Exhibit 10.6 shows an example from the leveraged loan market, where there are very few third-lien loans or loans priced below 60.

Exhibit 10.6 Examples of sectors with very small weights in a generic leveraged loan index

Seniority	Weight	Count
First lien	94.8%	1,184
Second lien	5.0%	150
Third lien	0.2%	3
Price Band	**Weight**	**Count**
Above 100	48.7%	647
90 to 100	45.6%	624
80 to 90	1.7%	21
60 to 80	3.7%	40
Below 60	0.3%	5

11

Other Topics in Corporate Bond Indexes

New Issues

Should new issues be immediately included in an index, as they begin trading, or should they be "seasoned" and included only in the next rebalancing period, for example at the start of their first full month of trading? In many investment-grade bond indexes, new issues are seasoned by adding them to the index only at the beginning of their first full calendar month of trading. This eliminates the problem of rebalancing the index midmonth on the date the new issue prices in the market. All market participants can depend on a stable set of constituents for the index for each month.

But in the high-yield bond and leveraged loan markets, one of the desirable features of new issues is that they are priced to trade up in price in their first few days of trading. This is part of the incentive to buyers to accept the risk of the deal. Of course, this price bump does not always occur with every new issue, but it is commonplace, and it does generate demand. Managers will push to receive their desired allocation to a new issue in part to register this price appreciation in their overall return for the month. Because this is often considered an idiosyncratic risk, it adds to the manager's alpha, which is an important consideration in the fees the manager can charge.

High-yield and leveraged loan indexes should capture this risk as well. That is not universally true for all of these indexes, but many do include new issues starting at their issuance date. If new issues in these markets are added after seasoning, the index will lag in total return due to the absence of new issues in their first month of trading.

Defaults

Should defaulted securities be included in below-investment-grade indexes? Like new issues, defaults are held in managed funds, and high-yield and loan indexes should include them to capture the same risks. But including defaults impacts the calculation methodology for an index. By the usual definition, defaulted securities pay no coupon, so any measure that is dependent on coupon cash flow can no longer be calculated. This includes yield, spread, duration, and the coupon interest component of total return.

In an index in which defaults remain in the index after defaulting, the averages of measures that depend on cash flow (yield, spread, etc.) must exclude the defaulted issues. This means that there must be a calculation of market weight using the subset of the index that includes only the non-defaulted issues because it is that weight that is used in the average for these calculations.

For total return, all issues are included, including the defaulted issues, but because there is no coupon interest, after the default only the security's price change contributes to total return.

There is another adjustment that can be made for defaulted securities. A security that defaults has been accruing interest up until the date of default, and that interest has been included in the total return for the security. At the date of default, that accrued interest should be subtracted from the total return in a one-time adjustment. In effect, this generates a negative interest component to return for one month for the defaulted security.

In many cases, a defaulted security is trading at a steep discount, so its market weight in the index is relatively low. For that reason, this negative interest return has only a small impact on the overall return of the index, and you could argue that it can be neglected. But because portfolio managers must make a parallel adjustment to their funds' net asset value (NAV) when a default occurs, subtracting the accrued interest they will now not receive, it is only proper that the index make a similar adjustment. Further, during periods with high default rates, the amount of the never-paid accrued interest from defaulted securities can be relatively large, so the adjustment to total return can be material to the index results.

Issuer Size

Should all issues in the market be included in a broad-based index? In markets with good transparency and liquidity, it is possible to include nearly every security in the market in the index. This type of construction can lead to a size bias in the index, especially when the weights and results are examined by sector. If the market has a few issuers with a very large total face value of debt outstanding in the market, those issuers can have an idiosyncratic effect on the entire index, as well as on the sectors into which those issues fall within the index.

Consider the case when General Motors Corp. and Ford Motor Co. were downgraded in 2005 and became high-yield-rated issuers. At that time, the U.S. dollar high-yield market exceeded $800 billion in market value. GM and Ford had many debt instruments on their books, but using a rule common to many high-yield indexes, which include so-called "straight" debt, the weight in the market of GM and Ford's bonds exceeded 12% of the market value of the entire high-yield market. Within the automobile sector of the high-yield market, these two companies exceeded 80% of the total market value of the sector.

In this example, for a high-yield index that included every issue, GM and Ford would have a disproportionate effect on the index results, representing over 12% of the market weight in a market with over 1,000 issuers. For the automobile sector of that index, the two companies would dominate the results with more than 80% of the market weight of a sector, which had about 50 companies at the time. The results of the auto sector of that index would reflect almost entirely only the idiosyncratic risk of these two companies. These concentrations would violate most rules for any diversified managed fund.

A solution is to constrain the exposure of very large issuers. This can be done in several ways.

One method is to impose a cap on the largest size an issuer can compose in the index. For example, if you impose a 2% cap, any issuer that has a market weight above 2% has its bond's face value reduced pro rata until the weight is 2%, and all of the other issues in the index are increased pro rata by the same amount, so the overall market value of the index remains constant. This method is attractive when the index is used as a benchmark because a 2% limit is a reasonable upper bound for an issuer in a diversified portfolio. In the case of GM and Ford in 2005, their weight in such an index would be 4%, instead of over 12%. Given the market value of the automobile sector at the time, GM and Ford would compose about 25% of the weight of the sector, rather than 80%.

Another method is to constrain the number of issues permitted in the index per issuer. This might not lead to an absolute ceiling for the market weight of an issuer, but it also doesn't require adjusting the face value of each issue in the index. For example, the index can include the two largest issues of each issuer, or one bond from each seniority level of the issuer. This method still captures the idiosyncratic risk of every issuer, but it reflects the real par values of the issues included.

With an index that includes the two largest issues of each issuer, GM and Ford would constitute a little over 3% of the market weight of the index and about 50% of the market weight of the automobile sector.

The example of the downgrade of GM and Ford illustrates that any selection criteria for an index imposes its own biases. There is no selection criterion that will weight all of the risks in line with any investment strategy. At best, an index can only approximate a range of investment strategies for a market.

Liquid Indexes

What is a liquid bond index and how is it used? Some markets have a mix of both liquid and illiquid securities. The selection criteria of a liquid index specify that the index holds issues that are considered most liquid and trade frequently. Such an index is usually constructed by a committee of traders from all of the major dealers in the market in order to get broad agreement on which issues are considered liquid. The issues in the index are rebalanced periodically, usually several times a year, because some issues may trade frequently for a period, but then become less liquid while others begin to trade more frequently.

Liquid indexes are a useful tool for managers whose strategy is to actively trade in the most liquid securities in a market, such as hedge fund managers. They are also used by the dealers who want to compare the P&L of their trading books with the most liquid issues as represented by a liquid index. Liquid indexes can provide insight into how that portion of the market is behaving.

However, one of the advantages of holding illiquid securities is the "liquidity" premium. Illiquid securities cost more to trade, so, all else being equal, they tend to trade at a lower price, and wider spread, than a comparable liquid security. Managers who buy illiquid securities are attracted to them, at least in part, because of this premium.

Liquid indexes do not hold these securities and can provide no information on that part of the market.

Investable Indexes

Can an index be sold as a product? Some investors want to make an unmanaged investment in a market as a whole, so a broad-based index would appear to be an attractive investment vehicle for such an investment. In addition to being a "neutral" investment in the market, the investor also stands to benefit by lower fees because there is no active manager and team of analysts to be paid. Finally, the investor wants the ability to exit the investment without many constraints on liquidity of the investment.

Such an investor typically doesn't want to buy all of the securities in the index directly. Instead, a dealer will structure an investment vehicle that provides the economic returns of the index as a single instrument that the investor can buy.

In practice, turning an index into a product by creating this structure depends on the underlying liquidity of the securities in the index. The securities in the structure must closely track the performance of the index, but not every bond in the index may be available for purchase in the correct size to mimic the index exactly. Liquid indexes are a good solution to this problem, making it far more likely to be able to buy all of the bonds of the index to put in the structure.

In periods of volatility, there might not be a balance between buyers and sellers in the trading of the instrument that represents the index. The dealer is exposed to the risk that there might not be buyers of the index instrument when prices fall, and the securities in the structure will be left on the dealer's books at a much lower value than what the dealer paid. The fees that the dealer charges the investor must take this risk into account, perhaps making the fee as high as what the investor would pay on a managed portfolio.

There are solutions to some of these problems, via exchange-traded funds, total return swaps, credit default swaps, and other contracts. These are discussed in more detail later in the book.

Indexes Versus Portfolios

When a portfolio manager compares his or her performance and weightings with an index, he or she should be aware of the many differences between the two, most of which were touched upon earlier in this section (Section IV, “Indexes”) of the book:

- An index will invariably have more holdings than any portfolio.
- An index will have weightings in some sectors of a market that the portfolio manager may choose to ignore, such as the gaming industry or defaulted securities.
- A broad-based index is not concerned with trading liquidity, unlike a portfolio manager.
- An index is not concerned with transaction costs or influencing market prices when buying or selling a bond or loan.

Closing Comments on Section IV

Indexes are one of the most widely used tools for analytics in all security markets. In the corporate debt markets, this is true, too. However, the corporate debt market does not have a single, dominant index such as the S&P 500 or the Nikkei 100 in the equity markets.

Virtually all of the indexes that are widely used in the corporate debt markets are run by investment banks. You must be aware of the selection rules and policies of the ones you use so that you understand their differences, strengths, and shortcomings. These benchmarks can be central to driving much of the macroanalysis that you are likely to do on the market and the analytics that you will likely perform on your portfolios and trading positions.

Keep in mind that there is a difference between an index and a comprehensive database, and they serve different purposes. You will want to use both. An index provides a relative measure of how a market and its subsectors are balanced and perform. A database is a comprehensive list of all of the securities in the market you are analyzing, including prices and analytics. For example, a database can be useful for screening the market for issues that you may want to purchase for a portfolio. Unlike an index, a database might not be a good reflection of the market: Some securities might not be representative, some security prices might be stale or inaccurate, or characteristics of some securities might not be maintained as accurately as the ones in an index.

Section V
Analytics from Macro Market Data to Credit Selection

By Robert S. Kricheff

This section outlines some analytic techniques to derive macro investment themes all the way down to identifying specific investments that might fit such themes. First the section looks at some of the ways that work can be done to try to predict how an asset class may move or react in certain economic and financial environments. The analysis is generally based on historical data.

This type of analysis can focus on returns as well as changes in relative value and historic relationships. This type of analysis can often focus on how different segments of the market reacted during various cycles in the past, such as examining how asset classes performed during a growing economy, an oil shock, or a rise in interest rates.

After examining how asset classes might react in these different scenarios, you will likely want to explore a specific asset class and see how sectors of that asset class might react in different environments. For example, within the high-yield sector you might want to examine which subsectors of the market re-acted the best during a recession—perhaps exploring this by credit rating, coupon, industry group, or other categories.

After you have developed a theme about what the best sectors to invest in for a specific environment might be, you will want to use techniques to identify specific investments that are worth analyzing.

This section explores these three layers of analytics: first the macro comparisons of various asset classes, then exploring subsectors of a specific asset class, and finally trying to identify specific investments that fit the themes that have been developed.

12

Top-Down Basics—Looking for Investment Themes Between Markets

Market Comparisons—Returns

Comparing the returns of various asset classes is one of the most common items to look at across markets. The comparisons are usually run on a total return basis over a comparable period of time.

As you compare returns over time, you should take particular note of patterns through various cycles—perhaps highlighting how different asset classes reacted during specific economic periods.

Comparison of asset classes is valuable for several types of end users. An investor or asset allocator who can invest across some of these different asset classes would want to use this information to compare the relationships in the performance of various asset classes and decide how to allocate money. Additionally, an asset manager or head trader who oversees positions in just one asset class might want to analyze this information to get a sense of potential inflows or outflows in his own asset class. For example, if you managed an investment-grade trading desk, you would care if returns in the high-yield bond market were rising more than investment-grade returns and ask yourself, "Does this mean credit risks are declining and money is moving to riskier asset classes? Is high yield becoming more attractive on a relative basis and will it draw assets away from my market?"

These observations might lead to further analysis that might make you decide on the size of your positions in various assets and the level of risk you are willing to take on your trading desk.

A U.S. dollar investor might compare the following typical asset classes:

- U.S. ten-year treasuries
- U.S. Investment-Grade Index
- U.S. High-Yield Bond Index
- U.S. High-Yield Loan Index
- U.S. $ Emerging Market Debt Index
- Asset Backed Securities Index
- Dow Jones Industrial Average
- S&P 500 Stock Index

Volatility

Investors understand that various levels of returns typically come with varying degrees of risk. So along with analyzing the returns for various asset classes you will want to analyze the associated risks. These risks are usually measured in terms of the volatility of returns. Typically, standard deviation of returns is used to measure the volatility. This is often then utilized in the Sharpe ratio, which was explained in Chapter 4, "Terms," and measures the return per unit of risk.

Different investors are willing to tolerate various types of volatility and might be drawn to certain asset classes or subsets of an asset class and they will often use the Sharpe ratio to compare their choices. The Sharpe ratio was originally designed to be used in the equity markets, but it is now widely used in the debt markets as well. The fixed income and maturity of debt structures tend to lead to lower volatility for these types of assets versus equities.

Correlations

Correlations of the returns among various markets can often be critical in determining which asset classes to allocate money to. The correlation of returns among different asset classes can be done using r. This allows investors to get a sense of how diversified their investments truly are. For example, assume an investor put money into two different markets hoping to get diversification. If the returns of the two markets are highly correlated through many different cycles and over a long time period, the investor may recognize that he or she might not truly be getting the diversification that he or she sought.

An asset manager might also be interested if correlations between certain assets increase or decrease during certain economic cycles. Perhaps an investor sees correlations of returns of all fixed-income asset classes increase dramatically during periods of lower interest rates. This might influence him or her to pool more assets into one asset class during these cycles.

In striving to find relationships that might give an investor or asset allocator an indication of upcoming market movement, you might want to run leading and lagging correlations of returns, too. Investors often do this with credit default swap indexes known as CDX indexes or iTRAXX, which are discussed later in Chapter 18, "Credit Default Swaps and Indexes." These leading and lagging correlations should help show you if there are relationships between assets where one moves ahead of the other. If relationships are strong, investors might use this to flag when weightings could be shifted from one asset to another. For example, does a drop in the return of small capitalization equities signal a coming drop in high-yield bonds?

Graphically looking at performance of various asset classes can be very helpful, too. A graph of returns for two markets is shown in Exhibit 12.1. One looks at the data on monthly returns, whereas the other shows a different look and uses the same data but shows it as $1 invested in each market since the beginning of the period. The latter shows a much more dramatic difference.

Exhibit 12.1 Comparison of returns shown as monthly returns and as if $1 were invested

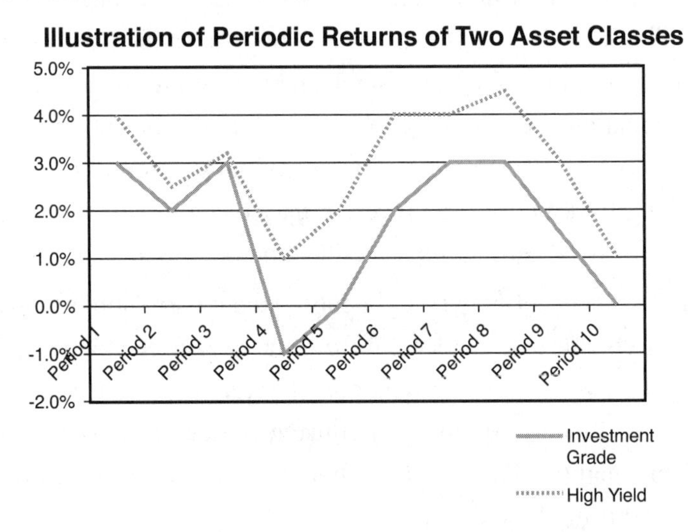

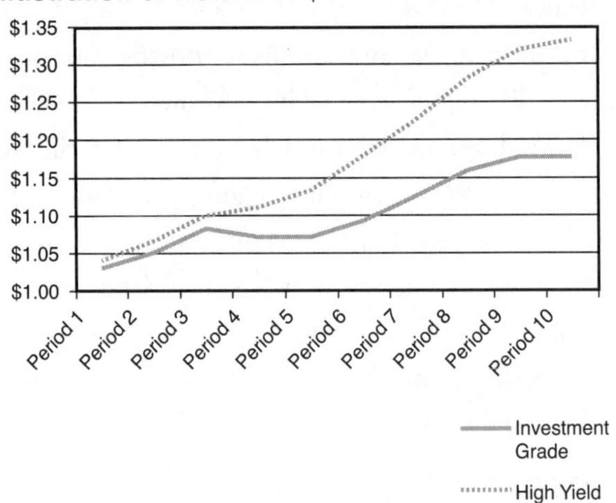

The analysis in Exhibit 12.1 focuses on total return, which encompasses the combination of interest and principal gains and losses. It

can also be interesting to break out the components of the return to look at the principal movements alone. This works best when comparing fixed-income asset classes, breaking out principal and interest, but with equities, dividends can sometimes be a meaningful component of returns as well and are worth breaking out. Showing just the principal changes for a fixed-income market helps show any volatility more dramatically. When comparing various markets or sectors and looking at just principal movements against the comparable total returns, it can highlight how much just the coupon or just the principal is impacting the return and how well the average coupon is compensating you for that volatility. For example, if one market had an average ten-year maturity and an average 10% cash-paying coupon and during a year that market had gone down an average of 5 points from par to 95, it has destroyed or used up half of the coupon and, in fact, has underperformed a market with an average ten-year maturity and average 7% coupon that has dropped on average 1 point in the year from par to 99. In this case, assuming both markets are in the same currency, it is likely that during the year the prices changed, not due to major interest rate changes but to a change in the perception of risk between the two markets.

Market Comparisons—Relative Value

Not only do you need to compare total returns when looking at various fixed-income markets, but you also need to compare bond data, including average yield, average spread, duration, and so on.

Yields and spreads are often used to measure relative value between various markets and subsectors of the market. For example, a comparison of yields over time of various markets might show that the relationship in yields between emerging market corporate debt and U.S. high-yield corporate debt has gotten to the tightest point it has been in the last several years. This might be an indication that the trend is about to reverse and this spread is going to widen again. If you are an investor who can invest across both asset classes, you might want to move money out of the emerging market debt asset class and into the U.S. high-yield debt asset class. Or if you believe this decline in spread is part of a very long-term, gradual trend, you might want to explore if the differences in risk and volatility between the two asset classes have changed materially enough to warrant this tighter-than-average relationship being sustained for a longer time period or even tightening more.

Dispersion

Dispersion of spreads and yields is not examined as frequently as some of the other items outlined previously, but can prove to be very insightful. You can look at the dispersion over time of yields and spreads of various markets. Dispersion of yields and spreads can be an indication of how much a market is differentiating for credit quality. During a period of very limited dispersion, it is implying that the market is not compensating investors for risk as much as in other cycles (or that risk is actually lower), and it might indicate that it is worth moving into the higher-quality tiers of the market or investing in another fixed-income market where perhaps the dispersion is greater.

Duration

As discussed in Chapter 4, *duration* is effectively a measure of how sensitive the price of a bond is to changes in interest rates. It is a good measure of potential volatility. Average duration for various asset classes can be compared to show which one you might want to favor in an increasing or decreasing interest rate marketplace. Note that duration is typically run on a yield-to-worst basis. Suppose a bond is trading at a premium and is trading to an early call date. If the price goes down enough, it might extend to trade to a longer date and see an increase in its duration; an effective option-adjusted duration tries to capture the impact of these embedded call options.

It is also worth noting that duration is typically assumed to relate to a general change in the interest rates, but if a market, an individual sector of the market, or a specific bond experiences a credit event that changes its credit quality and where it will trade, this is effectively repricing the interest rate for that sector of the market. In addition, if the credit quality of two market sectors both improve the same amount, the one with the longer duration will see a bigger price movement, just as if the general interest rate structure in the economy had changed.

Some Other Comments—On Using Historical Data

Like much of the work in corporate debt analytics, the top-down type of analysis can be a valuable tool for many market participants, but it can also produce several traps that you can fall into. Most notable is that analytical work is usually dependent on historical data, but the historical data might not be a proper reflection of the current market.

The makeup of the debt markets changes over time; because of the maturity structure of debt, they naturally change more than the equity markets. Changes in the makeup of the markets can be especially pronounced in the leveraged and emerging debt market segments because the companies that issue in these markets tend to be going through greater changes and the fact that these markets have been growing more rapidly than investment-grade markets also makes them more dynamic.

Ignoring how the constituents of a given asset class or subset of an asset class have changed over time can lead you to misleading conclusions from the data. For example, suppose within the high-yield market you want to compare how high coupon bonds performed during the interest rate increases in 1994 and 2004. The bonds performed much better in 2004. To analyze this, you might start to look at the industry groups, or what was going on in the economy, but one major factor was likely that the composition of "high coupon" debt issues changed dramatically in those ten years. In 1994, there was a fairly high percentage of deferred pay bonds outstanding (zero/fix or Pay-in-Kind, "PIK") that made up the universe of high coupon issues—these were more volatile and typically very junior in the capital structure and had longer duration than a typical high coupon cash-paying bond. However, in 2004, there were very few deferred pay issues. Therefore, you could argue that even the higher-coupon paper in 2004 was less risky than it had been in 1994 (at least on a structural and volatility basis) because the issues in the universe changed.

Some Other Comments—Weightings

Typically when you are comparing the data on different asset classes, you are using data from different indexes, perhaps the S&P 500 and the Credit Suisse U.S. High-Yield Index. This is a time to be aware of the differences in how some of these items might be constructed. One of the things to be cautious of is how weightings are used.

Most index data is market weighted. This means that if you are looking at yields for the U.S. dollar high-yield bond market index versus a U.S. dollar emerging market index, the yield of each bond in the index would be factored by the amount outstanding on the bond before being included in the calculation so that the yield on a $1 billion issue would have a bigger impact on the figures than the yield of a $500 million issue. Additionally, if market weightings are used, each of the issue sizes would be multiplied by its recent price (which is the most common way to do it). This creates some obvious biases that you should be very cognizant of, particularly when you are looking at slices of the market and subsectors—a sector that is seeing rising prices increases its impact on the overall market's results.

Be aware that some statistics that you are looking at might not be market weighted, whereas other data is market weighted. For example, suppose you are comparing an emerging market index to a U.S. high-yield bond index. An index provider will likely give you returns, yields, spreads, and duration on a market-weighted basis, but when asked the average coupon or the average maturity, that may be a straight average, which could impact conclusions of your work.

13

The Next Layer—Analyzing a Market

Introduction

All of the tools that were used to compare various asset classes in the prior chapter can be applied to the analysis of trends and relative value within a given asset class, too. You can subdivide an asset class in numerous ways so that you can develop themes about which segments of the market are either driving returns or increasing or decreasing volatility and also analyze where there are relative value misalignments. This analysis should help you to develop investment themes and strategies and observe trends that are evolving in the markets. Or if you are involved in more of an investment banking or capital markets role, it might enlighten you about the types of transactions that would be attractive to bring to market.

Total returns for various periods can be compared to show what has driven performance or what areas of the market outperformed or underperformed in various environments.

Some of the typical types of analysis that can be done include the following:

- The volatility of returns can be examined using standard deviation.
- Relative value of various market tiers can be explored through comparing yields and spreads.

- Correlations of returns between various tiers can be examined.
- Duration of various tiers in the market can be compared.

These are some of the typical types of comparisons that are run on the various market subdivisions. The following sections explore some typical segmentation that is used to examine the various corporate debt markets.

Risk Segments

Two typical ways to segment the market by levels of risk are to look at the various rating tiers of the market or to look at the market segmented by bond rankings.

Rating categories can often be used as a general proxy for credit quality within a market. So, although there are many shortcomings of the ratings, they are good for this general view of risk within a marketplace and are, therefore, a logical way to slice the market up when looking at credit risk differentiation (e.g., BB/Ba, BB/B, B/B, B/Caa, CCC/Caa).

These categories can be examined by returns, volatility, and relative value measures. You can also explore what the weightings of each category are in the marketplace to judge each tier's overall impact on the market.

When riskier, lower-rated categories are outperforming, it is often a sign that the market participants are willing to take more risk with their investments. When two different rating categories are trading tighter than historical averages, it might be telling you that you are not getting paid enough for the risk of being in the lower-rated tier.

Using correlations of returns over a long time period, you could see that more highly rated tiers are more closely correlated with treasury rates and begin to examine how this might be hedged, or simply look to avoid those tiers of the market because you want to minimize interest rate exposure and move to lower-rated issues. However, do not ignore that although you might be reducing the interest rate risk, this type of strategy will increase your credit risk in this scenario.

Bond-ranking categories are another way of comparing risk. When bonds that rank more senior, such as Senior Secured bonds, are outperforming, it is generally perceived that the market is avoiding risk and when the trend is reversed, the perception is reversed as well.

Similar trends can be detected by exploring the relationship between spreads and yields of the various ranking categories within a market.

However, remember that rankings can be tricky. There are often details in the description of notes that cause subtle changes in the actual way a ranking is treated for different events and there are details in any securitization agreement that involve numerous nuances. Understanding the details and structures can be key to not misinterpreting the analysis of rankings.

Bucketing Sectors

Although risk categories are naturally defined, you can create other categories that can shed considerable light on performance and relative value. These buckets can be used to explore any number of aspects of the market. The following paragraphs discuss some common buckets that are built and examined to get a sense of the trends that might be evolving in a market.

Maturity buckets can be easily designed. Let us look at the high-yield bond market; this market rarely has maturities of over 10 years. You could subdivide the market into the following segments: Maturities of 0–2 years, 2–4 years, 4–6 years, 6–8 years, 8–10 years, and 10+ years. You could then get all of the data for each category, including returns, average price, yield, spread, market weighting, run total return, volatility, and relative value measures and examine these factors through various cycles. Buckets are also typically done for duration, coupon, and average price.

Keep in mind that constituents in the buckets will shift over time, as maturities come closer or prices move up and down. Therefore, you must be consistent in using buckets formed at the same time to make any analysis valuable—usually the buckets are based on the data at the beginning of the time period you are examining.

You can see that a database has to be fairly robust and well designed to be able to build these buckets and to be able to change them over time. You have to make sure that all of the key data points that you want to use are identified in fields, that the fields are usable and sortable on all debt instruments entered into the database, and that the user interface lets you define universes from these fields.

Industry Analysis

Analysis of industry segments is widely used in the corporate debt market and often follows many of the patterns that you can see in the equity markets and utilizes some of the equity market tools. Industry rotations and weightings are one of the most common macro discussions in the corporate bond markets (and in the equity markets). For example, does a portfolio manager want to overweight metals and mining companies, or does an investment bank's trading desk want to increase its exposure to home-building companies?

Industry analysis, like other categories, can compare returns, yields, spreads, duration, correlation, dispersion of yields, and similar attributes. Sometimes the industry categories are grouped into larger buckets, such as cyclical and defensive, although energy is usually put into a separate category. These large buckets can give a big picture of what trends may be occurring or the risk profile of a portfolio more easily than data that includes 20–30 different industries.

Note that various market participants and indexes often subdivide industry groups differently. So when doing this analysis, make sure that you are comparing apples to apples, meaning that you are defining the industry categories the same as the benchmark that you are using.

Surprisingly, comparative credit statistics by industry are often ignored at this level of analysis. I believe this happens because credit analysts tend to work on one or two industries and don't look across sectors as much and also because different industries have different key characteristics and the relative value of credit metrics for different industries are not always easily compared. Also most databases that run bond and loan metrics often are separate from the databases where the analysts store credit statistics and metrics.

However, key financial figures and ratios can be normalized across industries and then more easily compared. It is exceptionally helpful to be able to have average key credit metrics by industry and would add greatly to the type of analysis that you could undertake. Some of the common types of items that lend themselves to being compared across industries could include the following:

- EBITDA margins
- Comparison of net free cash flow as a percentage of debt
- EBITDA coverage of interest expense
- Leverage ratios if they are adjusted for industry asset values[1]

Building Industry Equity Monitors

As an addition to industry analytics, it can be very valuable to build baskets of public equities by industry and to use these as monitors for changes in price movements and also as a valuation tool for asset coverage. Analysis of any leading and lagging relationships of returns between equity securities in an industry and their fixed-income counterparts can help trigger alerts for trading opportunities.

Equity valuation changes by industry should be monitored because they are a proxy for the value of the asset value underlying the debt. Additionally, strong equity valuations may supply companies within an industry the ability to issue equity as a deleveraging financing source.

When monitoring the equity valuations of an industry, most commonly the ratio that is used is the Enterprise Value to EBITDA ratio (EV/EBITDA), which can be used across all industries with public equity and positive EBITDA to give a relative value tool for companies within the industry. The enterprise value is the stock price of the company multiplied by the number of shares plus the net debt of the company.

You might want to build these equity industry tools in various ways. You might only want to include the equities of companies that also have debt outstanding in the universe that you are exploring. You might want to include all of the public companies in an industry or build your own equity comparable index. So, if you are building an equity multiple model for the media sector in the United States, you might have a dozen companies with high-yield bonds and public equities outstanding to include.

You also have to decide if you are going to market weight each company in your comparable equity universe or even weight them. In this case, the latter is often a better measure of typical company valuations in the sector.

What Can Be Learned from Market Shocks

It can also be very valuable to examine how various segments of a market performed during specific market shocks. These shocks might come from a war, an oil shock, or a natural disaster. Of course, by definition, a shock is typically not predictable, but the analysis is helpful to have so you can react quickly when such events do occur.

With limited trading liquidity, relative to stocks or government bonds, when shocks do occur, prices in the corporate debt markets can often gap down or gap up with only modest volumes until levels where trades can happen can be found. Also with little timely pricing data and many securities not trading regularly, the data that you can get about price movements around the time of a market shock can have significant noise in it or can be outright misleading. Therefore, it is best to examine prices for some time on either side of the market event. Even with these caveats, this type of analysis can be very valuable.

The Crowded Trade

Data analytics on much of the markets and market segments is more and more widely used. Many of the common screens and sorts are regularly run and published by sell-side firms for a wide client base. Similar systems are also widely used throughout the markets to do the sort screens and relative value comparisons discussed previously. The increased usage of these data analytics has created the phenomena of the "crowded trade."

For example, assume that the relative value as measured by spread to treasury between BB-rated issues and single, B-rated issues has moved to its tightest level in the past ten years and that economic news has started to show some weakness. All of this comes up on investors' and traders' screens and they all want to move out of one part of the market into another. If large pools of assets are trying to do this trade, the transactions might get harder and harder to undertake and prices of the two asset classes will "correct" or "overcorrect." This trade idea then becomes a "crowded trade."

This crowded trade has become more common in recent times in the equity markets, too, with electronic trading systems. It can cause much longer and sustained increases in volatility and disruption in the market when it happens in a much-less-liquid market like the one for corporate debt.

Some of the products that have been developed with the help of data analytics also have heightened the crowded trade and increased volatility in the corporate debt markets. In particular, the increased use of index and index-like products such as CDX and ETFs can increase the severity and volatility of these crowded trades in the fixed-income markets.

Adding to the crowded trade is that much of the macro work comes from the sell-side firms and is widely distributed. Internally developed work, early recognition, and speed in reacting are important in getting ahead of these trends. Do not be afraid to be early in identifying a trend. As General George S. Patton said, "If everyone is thinking alike, then somebody isn't thinking." Being early is critical in being ahead of the crowded trade, especially in the corporate debt markets because the limited market liquidity makes it difficult to move positions too quickly, so early moves are good.

Endnote

1. Industry asset value could be based on comparable equity multiples or recent acquisition multiples and then used as a numerator over the leverage ratio to show a proxy for relative asset value coverage that would be more normalized across various industries.

14

Data Analytics for Credit Selection

Introduction

Once you have established some of the key trends of various segments of the market and this has evolved into investment themes, you need to be able to find lists of bonds and/or loans that fit your newly derived criteria. You also need to create comparisons of key data for these credits so that analytical work can be done on the lists and issues can be selected that you and your credit analysts want to delve into more deeply before you make the final decision to purchase or sell certain debt instruments.

Data analysis can help identify the issuers and issues of debt that you might want to own in your portfolio or on your trading desk, or, more important, it can help identify those debt instruments that you do *not* want to own. This can save considerable time guiding analysts, traders, and portfolio managers to be more efficient. Data analytics is also very valuable for establishing relative value among various investment choices during the analysis process.

Keep in mind that some of the important items that go into the final credit decision are not easily put into a format that is easy to quantify. These types of items include management meetings, site visits, and discussions with competitors as well as a company's competitive environment and the quality of a debt instrument's covenants. You can try to incorporate some of these items into a relative value scoring system for your investment analysis, but they will be significantly more subjective than items such as a revenue or operating margin number.

The goal of the credit selection criteria could be as simple as developing a diversified portfolio, with or without certain industry biases. Or you might want to find something more specific, such as CCC-rated issues with long duration, low dollar prices, and the best asset protection. You will find that much of this work in using analytics for credit selection tends to involve queries and sorts, and it benefits from having a robust number of fields to choose from in defining your search from a large database. Having a full complement of possible debt issues in the database is also critical to the success of the analytics.

Data for Credit Selection

Ideally, you would like to have significant amounts of data in your sortable database for each debt issue and issuer. In addition, of course, accurate pricing data is one of the biggest concerns for data on corporate debt.

Ideally, you want one database where you can search and sort data thoroughly on the debt description, pricing, yield, spread, and duration measures, but also include detailed issuer credit metrics. However, I do not know of any databases that contain all of the data you would want to include for credit selection. Think of the data for which you need to make credit selections and divide it into three categories: (1) data on the description of the debt issue and issuer (e.g., coupon, maturity, issuer, ranking, currency, covenants, industry, and country), (2) relative value tools to judge the debt issue (e.g., price, yields, spreads, and duration), and (3) credit metrics about the issuer (e.g., revenue, operating margins, and leverage ratios). Most widely used databases are strong on the first two categories (except for covenants) but particularly weak on the third category.

Over time, you will find that the number of items that you want to be able to run queries and sorts on is exceptionally large. The following is a list of just a few of the most important fields you would want to include on every debt instrument and be able to query by:

- Coupon
- Ranking
- Price
- Yields (including YTW, YTM, and CY)
- Spread (including spread for both YTW and YTM)
- Maturity (possibly including information on call structures, including clawbacks and special calls)
- Duration, option adjusted duration, and convexity
- Credit rating (ideally including the rating by the major agencies—S&P, Moody's, and Fitch—or sometimes another local country rating agency)
- Country of business (bond description should include currency and country where the debt is issued; these might vary from each other and vary from where the company does its core business)
- Industry segment (established industry segments are needed; you might want another code for whether the industry is cyclical or defensive, or consumer or industrial, etc.)
- Basic credit ratios (e.g., leverage, interest coverage, net free cash flow/debt[1])
- Size of revenue and capitalization
- Financial trends (e.g., recent revenue and EBITDA growth, operating margins)

- Enterprise valuations (e.g., public Enterprise Value[2]/EV for the issuer or comparables)
- Upcoming debt maturities
- Maintenance covenant headroom[3]
- Ownership structure (public equity, privately held, leveraged buyout "LBO")

The inclusion of all of these types of data, and much more, is going to be important in credit selection.

Comments about Sorts and Queries

Sorts and queries are ways to pull up lists of bonds and loans from your databases, determined by selecting a precise set of criteria. Hopefully, these lists that you develop build on certain investment themes that you have developed from some of the macro work that you have done.

The database needs strong multiple sort and query functions. This means lots of options of how to put together requests and how to extract the data records from the database for the end user (viewing each debt instrument as a record) and how to sort the data and display it. You want to have flexibility in your queries and be able to include multiple choices. For example, you do not just want to put in a request for a yield to worst between 5% and 11%. You want to be able to request a query for a list of bonds with that range, along with, for example, a maturity range, a duration range, a bond-ranking restriction, and a rating limit. Then when you get the output of this data with a list of bonds, you want to be able to sort it by any number of identifiers or fields as well, whether that is in some proprietary database or as simple as shifting it to Microsoft Excel and using its rather robust sorting functions within its Data tab. Exhibit 14.1 shows what a typical very simple tab might look like where you can request a query using certain criteria.

Exhibit 14.1 An example of a query screen

Sample Interface Screen for Query Function						
Industry	**Coupon**	**Rank**	**Rating**	**YTW**	**Spread**	**Duration**
All	All	All	BBB>X>D	X > 5.5%	X > 350	X< 5.0

As you develop your lists of possible investments, the team will need to analyze some of the key criteria of each security and its underlying credit. The rest of this chapter outlines some of the key types of metrics and types of analysis that are run to select which issues best fit the investment theme you are looking to meet as well as the best-quality investment for the criteria under which you operate.

An Example

Suppose you are a trader at a sell-side desk and the analyst who works with you expressed a view that the food industry is likely to post outperformance. She points out that you trade bonds in this sector much less than other sectors of similar size and that according to TRACE data there was considerable volume in these issues. Together you decide you should have more positions and become more active in trading the sector. Together, you might request your strategists, or whoever runs your database, to run a query that will list the five largest high-yield issues in the food industry, with key financial ratios linked to the analyst's spreadsheets; make sure the query includes prices, yields, and spreads of each debt instrument. Exhibit 14.2 is an example of what a simple query such as this might produce. A quick look at the data might lead you and the analyst toward taking a position in Mustard Corp. bonds because it has the highest spread per leverage ratio as seen in the far right column.

Exhibit 14.2 Illustration of data output for a query on food credits

Company Name	Description	Maturity	Price	Yield	Spread	Leverage	Spread/Leverage
Ketchup Co.	6.125% Senior Notes	2021	101	5.95%	356	3.1x	114.8
Mustard Corp.	8.25% Subordinated Notes	2020	101	8.30%	691	4.9x	141.0
Salsa, Inc.	9.125% Senior Notes	2019	103	8.21%	695	6.0x	115.8
Mayonnaise Co.	7% Senior Secured Notes	2020	101	6.80%	465	4.7x	100.0
Soy Sauce, Inc.	7.25% Senior Notes	2020	102	6.77%	504	4.7x	107.2

In reality, for this small of a sort, the analyst would probably do it herself, but if it were a much larger industry, the data systems would be helpful.

Financial Metrics

Notice that Exhibit 14.2 includes data on the securities and pricing and yield information that can be used in relative value. It only includes one credit metric. Ideally, the next layer of information that you would add from your database would include more detailed financial metrics.

Exhibit 14.2 does include what is probably the most widely used metric in the corporate debt markets, the leverage ratio. Although numerous variations exist, in Exhibit 14.2 we used the basic ratio of total debt/EBITDA. This ratio is a proxy for how much debt is on the cash flow of the company and is a very simple snapshot of how well the debt that you are potentially investing in is covered by the asset value of the company's cash flow.

When examining financial metrics, credits are typically first compared within their own industry, or at least closely aligned industries. This is due in part to the fact that a steel manufacturing plant is going to have different operating data and different business influences than an operator of television stations. Not only will the income statement results look different, but reinvestment rates and optimal balance sheet structures will vary greatly as well. However, even though comparisons of credit data are usually first done within an industry, keep in mind that eventually a credit decision is not just made relative to other opportunities in an industry, but should always also be made across the whole spectrum of your given investment universe.

Let's look at an example of why intra-industry comparisons are often done first. If the average public company that owns television stations is valued in the stock market at 10× its EBITDA, and the typical television operator has debt leverage of 5×, this would imply it has 2× asset protection. Another industry such as a steel manufacturer might have a lower valuation in the equity markets such as 7.5×; if a steel company had the same leverage as the television company, it would imply a lower asset protection of 1.5×. Numerous other characteristics make each industry have unique metrics and business cycles.

Industry groups are definitely not the only segments of the market by which you will run data to make credit selections. At different times and for various projects, you are likely to run data comparisons on debt in the same rating category or with comparable maturity or duration as you try to choose the best issues that align with a trend you expect to occur or to rebalance a portfolio.

Let's look at another aspect of displaying the data, as shown in Exhibit 14.3. If a company has various pieces of differently ranked debt, you need the metrics to be specific for each tier so that you can help distinguish its value. So if different tiers of the capital structure are being examined, the database should show the leverage through each level of debt, including all of the debt that is more senior as well as total leverage for the company. So if a company has Senior Secured bank loans, Second Lien bonds, and Senior Subordinated Notes, you would want to know both the overall leverage and the leverage at each level of debt to be able to compare with the variously ranked bonds and inter- and intra-capital spreads. You need to be sure to include data fields for both debt-level ratios and total company-level ratios in case you are doing a comparison across all Senior Secured issues, regardless of industry. You also do not just need to include the debt at the issuing level because the total debt level impacts the overall credit quality of the company even at the most senior level; you need the total company debt field so that you can include this as a valuation metric.

Exhibit 14.3 A multitier capital structure with leverage ratios

Description	Maturity	Amt. Out ($MMs)	Price	Current Yield	YTW	Spread in Basis Points (bps)	Debt-Level Leverage	Total Leverage
Term loan +350bps,100bps flr	2018	300	99	4.55%	n.a.	n.a.	1.0x	4.7x
5.75% Senior Notes	2020	500	100	5.75%	5.73%	377	2.7x	4.7x
6.125% Senior Subordinated Notes	2021	250	101	6.06%	5.95%	355	3.5x	4.7x
Total Operating Company		1,050					3.5x	4.7x
Holding Company								
8.625% Senior Notes	2020	350	101	8.54%	8.30%	691	4.7x	4.7x

Notice that the leverage at the individual bond level goes up, while the total leverage remains constant, but we include this in every line so that there is an associated field with this data for every security in the capital structure.

Numerous other financial ratios can be compared and by which the companies can be sorted and ranked. Net free cash flow as a percentage of debt is a favorite ratio of mine to use. But there are also ratios for coverage of interest or coverage of interest after capital expenditures. In addition, you might want to analyze numerous other items. These might include working capital, return on invested capital, and so on. These types of items need to be placed in fields and regularly updated, but you have to balance what is realistic to include in a database and keep maintained, versus timeliness and manpower constraints.

Operational Data

Ideally, the databases that you are using can go beyond just the basic financial ratios and include some other operational financial data for comparisons as well. Some of the typical data might include trend analysis that could show recent changes in revenue and EBITDA. Growth rates could be based on the latest quarter results or could be shown on a rolling, four-quarter average. When this data is done across an industry, this should be able to show outperformers and laggards in either tabular or graphic form. In this example, before any conclusions were made, you would want to analyze if some companies grew through acquisitions and if some grew organically.

Other operational comparisons can also be included in databases that often prove to be very valuable. These might include various margins, such as gross profit margin, operating margin, or EBITDA margin. For some businesses, such as retailing, turnover ratios of inventory, payables, and receivables can be valuable to compare for operational differences.

Financial Liquidity and Some Differences Between Credit Analysis and Data Analytics

Financial liquidity refers to the cash and excess cash from operations that a company has after meeting all of its bills and debt obligations. It is a critical factor that can differentiate credits. Let us take a quick look at how an analyst doing detailed credit work would want to analyze this factor relative to how you might want to design it into a database for sorts and queries and comparative analysis.

The first level of liquidity that a credit analyst will likely look at is the amount of liquidity generated or used by operations. There are many nuances in how people prefer to examine this data; some prefer to do it before capital expenditures and debt repayments, some prefer examining it afterward, some prefer to net out dividends and stock buybacks, and others do not. Some simply use the data from the statement of cash flows in the financial statements; others choose to calculate it with their own adjustments (deciding among all these different choices of this calculation is a good example of the advantage of having your own analytical team versus just plucking credit data from a third-party server). However you calculate it, it is an important part of comparative analysis. To analyze this operationally generated liquidity, the credit analyst must pull items from both the statement of cash flows and the income statement in the financial statements of a company. Ideally, the analyst must determine what portion of items are discretionary capital expenditures, or one-time in nature versus required maintenance spending. The analyst also needs to analyze all of this data over time. For the record, I prefer deriving a net free cash flow number two ways. First, using operational sources only (as opposed to financing sources) net of everything but stock buybacks, dividends, and elective debt repayment and then also analyzing what is left after you net out these three items. The first one gives an idea of what they could generate without the elective corporate actions, whereas the second shows how the management chooses to spend cash away from operating the business.

But all of this data that the analyst needs to make a conclusion is not going to go into a database for sorts and queries. It is best to derive a simple ratio for database sorts and queries. A good liquidity ratio is net free cash flow as a percentage of debt. These ratios are particularly important for leveraged credits as the ability to pay down debt can lead to meaningful credit improvement. So, ideally, in the interest of space, you would build a field to show the credit ratio (net free cash flow/debt) and not show all the other levels of detail that go into the individual credit analysis.

Note that ratios or percentages (which are actually ratios) are usually best to use when running data metrics for credit selection because it helps to compare credits across a broad spectrum. However, when doing detailed credit analysis, you not only need to look at ratios, but also at many other items. For example, you do not want to completely ignore the sheer size of a company, as scale and size are sometimes considered valuable by analysts and decision makers and can make an economic difference for some companies.

Just as a point of interest, capital expenditures is an item that is often examined, too, as it is a use of cash that is not in the income statement and is not captured by EBITDA or typical adjusted EBITDA, but can materially change the credit profile of companies and industries. (Some industries are very "capital" intense and others are not, so just using EBITDA can be misleading.) Capital expenditures is a cash flow item that is often asked about in comparative metrics and can be shown as a percentage of revenue. Again, whether special, one-time projects (e.g., installing new machinery in a plant) are counted in these ratios needs to be determined by the credit analyst.

From the credit analyst's side, the analysis of liquidity does not end with the ratio that goes in the database. The credit analyst also needs to analyze total liquidity as a percentage of debt. Liquidity sources would typically include available bank borrowing, cash, and liquid securities held on the balance sheet. The credit analyst might want to determine if some of this liquidity is earmarked for specific spending (e.g., a capital project, an announced stock buyback, or upcoming maturity). Although difficult to quantify, an analyst also needs to consider if a company has access to the financial markets and if it can raise more debt or issue equity to help address liquidity. Digging down another layer, a credit analyst has to determine if there are other hard assets that the company could sell to raise cash and if these assets do or do not impact EBITDA and other cash-flow measures if they are sold.

Then there needs to be some measurement and analysis of upcoming debt maturities; liquidity from existing or new sources will be needed to service these maturing obligations. You should examine maturities at least coming due over the next 12 months, but I recommend looking out at least five years whenever possible. For comparison's sake, this can be put into a ratio as a percentage of total outstanding debt, or perhaps over the EBITDA level for the latest 12 months. This would be atypical to include in a database.

You can easily see that it would be difficult to include the level of analysis that a credit analyst needs to undertake to make a decision about a credit into a sortable and comparable database. The data on which these credits are analyzed is a small subset of the work that is done for final investment determination. However, the database work can save hours of time in determining where the analyst should focus and the comparative items that can be drawn from a database are critical in helping to make decisions.

Exhibit 14.4 is a graphic view of the liquidity ratios discussed, comparing a few companies over time.

Exhibit 14.4 Illustration of comparison of NFCF/debt

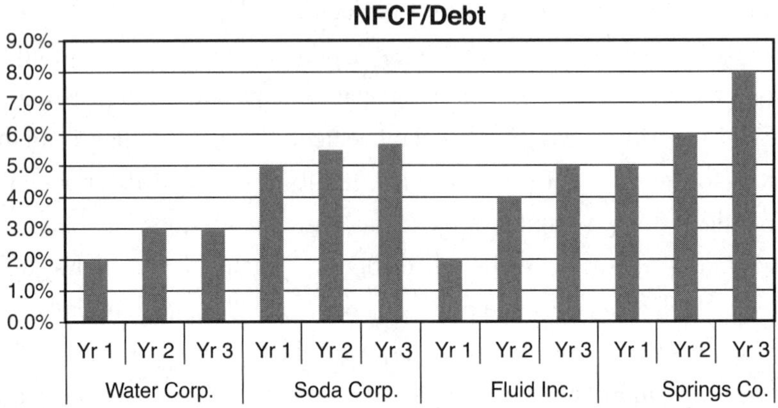

Credit Scoring

Some firms utilize credit-scoring systems. These range from very basic and relatively undisciplined methodologies to more detailed and complete methodologies. Other third-party analytical groups and some parts of investment banks often use scoring systems as well. Credit scoring is also effectively what the rating agencies try to do. Scoring systems can be extremely valuable to be included in an internal database for sorting and relative value analysis.

Investment banks use scoring tools, particularly for loan hold positions,[4] but there are other business considerations at play. I think it would be an improvement to see them increasingly utilized as a risk management tool on sell-side trading desks, too, though these positions are not typically held for periods as long as on the buy-side.

Credit-scoring systems vary depending on what the goal of the end user is. An insurance company, a mutual fund with high turnover, and a bank would all want different metrics measured and weighed differently in their scoring systems. In addition, although this book refers to it as a "credit" scoring system, it might also encompass trading and bond traits into a final score.

Scoring systems need to be flexible and use data that is readily available. They can be set up within an industry for ranking credits. However, a more-effective system is to have a fixed scoring system that is used across all industry groups so that comparisons can be run within and across industry groups.

Let us go through a very simplified version of how this could be run from a database to analyze the strength of a few different credits in the same industry. This example will give you an idea how it might work, but it is not the ideal way to run one. In the upcoming example in Exhibit 14.5, each ratio that you check is ranked for the credits from strongest to weakest, with the strongest getting a 1 and the weakest getting the highest number. You can do this for a number of ratios, and then you add up the scores, and rank the credits. In this system, lower numbers are better.

This is a very simplified scoring system. To make it more meaningful, you could add weighting to the different factors. Items that you deem as more vital could be more highly weighted than other factors. This generates a scoring system that forces ranking.

More often, systems are set up with absolute scores that can be used across industries and new credits can be added without impacting other credit scores that have already been calculated.

For example, assume that for an industry (in Exhibit 14.5, the demolition industry) you have established that an average enterprise value multiple is 6× EBITDA (this could be determined using the multiple at which equities of companies in the same industry trade or where recent acquisitions have taken place, or both).

You could then take the leverage ratio and see how well each company's debt is covered by this implied asset value of 6× EBITDA. If a company has debt equal to 3× its EBITDA (a leverage ratio of 3×), this implies that its asset value covers its debt 2× (6× ÷ 3× = 2×).

The ideal way to do this could be to set up buckets. For example, in a system where the low score is better, you could give:

1 point if the leverage ratio is 2× or lower (up to 3× theoretical asset protection 6× ÷ 2× = 3×)

2 points if it is 2.1× to 3× (up to 2× theoretical asset protection 6× ÷ 3× = 2×, etc.)

3 points for 3.1× to 4×

4 points for 4.1× to 5×

5 points for 5.1× to 6×

7 points for 6.1× and higher—notice the higher penalty for exceeding the asset value

This could be run across any number of ratios. A system such as this is more flexible because it can be used across various industries and sectors. One of the important factors to keep in mind is how differences in various industries can be adjusted to make comparisons fair.

Exhibit 14.5 Illustration of a ranked scoring system and a bucketed scoring system for one metric

	Leverage Ratio	Ranked Credit Score	Bucketed Credit Score
Demolition Corp.	2.00×	1	1
Knock It Down, Inc.	2.50×	2	2
Blown Up Corp.	2.75×	3	2
No More Bricks, Inc.	3.10×	4	3
To The Ground Co.	3.50×	5	3
Destroy, Inc.	3.50×	6	3
Clear the Way Corp.	3.80×	7	3
Building B Gone, Inc.	4.50×	8	4
Wrecking Ball Co.	5.10×	9	5
Piece by Piece Co.	5.75×	10	5
Average Industry Equity Valuation Multiple			6.00×

It is important to note that scoring can be set up for management history, corporate governance, ownership structures, and so on. These items obviously have a much more subjective aspect and various groups using scoring systems will have different views on which of these items to score and their relative importance.

Typically, numerous items are included and a total score is given within a scale (e.g., 0–100 points). Sometimes an "unacceptable" range is set in the scale. Depending on how the credit score is managed, the system might be able to address both a scoring system for the issuer and the specific issue. Remember, complete analysis incorporates the characteristics of the issuing credit and the bond or loan that is being examined.

These can be set up in numerous more complex ways, but this should give you a basic idea of how data can be organized and used in a credit-scoring system.

Credit-scoring systems are not the easiest thing to set up from scratch. Once they are set up and maintained and reviewed regularly, they are one of the easiest items to include in a database along with the debt instrument data. The credit score is valuable in these databases because it is a single field that addresses a broad swath of credit quality issues.

Analytics Used in Relative Value

Relative value is comparing the risk and reward of two or more investments and choosing which ones you would buy or sell. Much of this analysis in fixed income uses spread or yield as a measure of reward or return. As far as measuring risk, this section does not look at volatility of returns as a metric, but uses credit quality as a measure of risk.

Different investment strategies and types of investors have different tolerances for credit risk, different return parameters, and different time frames. There are also varying views as to which credit metrics best measure risk; plus some of the credit decisions may be based on more subjective scoring of credit considerations. Therefore, even with the most detailed analytics on a credit, different investors may reach different conclusions.

The key metrics used in relative value could be a series of credit ratios and measurements or it could be a final credit quality score, as discussed in the prior section. Ideally, the database will have the ability to run whichever combination of these is asked for.

As far as the measurement of return, it is critical that the bond and loan data has up-to-date prices, yields, and spreads. If these items are too far off base, they prevent the analysis from being meaningful.

As explained earlier, debt instruments have various ways of being valued. The simple price of the bond or the loan is usually not used as the key measure of value, unless it is highly distressed. Yields are used more often than price. To be conservative, the lowest yield is typically used (whether it is to maturity or a specific call date); this is typically called a *yield to worst (YTW)*. Yields become less helpful for comparative purposes if the debt that you are comparing has meaningfully different maturities from each other. As the yield curve is usually upward sloping (or higher yields are on longer-dated bonds), if you compare two bonds that are fairly similar in all aspects but maturity and only look at YTW, a bond with a ten-year maturity will almost always look cheaper (i.e., offer more yield) than a bond with a two-year maturity. So unless all of the bonds on which you are doing a relative value analysis have very similar maturities (and call schedules), the spread to worst is the better tool to use for basic relative value comparisons. (Remember, the *spread to worst* is the yield minus the yield on an equivalent benchmark bond. So, typically, on a U.S. dollar corporate bond with a five-year maturity, you would measure the spread by taking the yield on the corporate bond and subtracting the yield on a typical five-year U.S. Treasury Note.)

Putting these measures in the context of a risk return analysis, the yield or spread is the proxy for the return (in this case, the future return), whereas the credit measures are the proxy for the risk. So you will now put them into a ratio (similar to how the Sharpe ratio uses historic returns and the volatility of returns to measure risk reward).

Suppose you have decided to use the spread as a measure of return. The tighter the spread implies the better the market believes the credit quality of the debt issue is. You can take this and divide it by the credit scores and see what the spread per point of credit score is. The higher this ratio, the better value you are getting for the perceived credit risk. It could also be done using the leverage ratio or some other credit metric.

Exhibit 14.6 shows how the same industry comparison of credits might look different using a basic ratio analysis in the top section versus a credit scoring system in the bottom section of the table.

Note that this works well because in theory the lower spread and the lower credit score both imply a better debt issue. If you were using a ratio of EBITDA/interest coverage, in which a higher ratio implies less credit risk, or a credit-scoring system where higher numbers were better and wanted to bring these ratios in-line with the prior examples, it might be easiest to use the reciprocal of this interest coverage ratio or the credit score.

To reach an actual conclusion, you need to consider more factors than just one credit metric. Numerous other financial and structural issues enter into the decision; perhaps a lower dollar price is desirable or a greater weighting on the debt instrument's duration needs to be considered. But these basic relative value ratios can be a very valuable sorting tool.

Another method that has been used for relative value measures is a regression analysis. Using a regression does allow you to use a graphical representation and shows which securities are theoretically rich and cheap and the amount by which they are. But again, it is driven by which ratios and yield or spread measures you want to input into the regression analysis.

Exhibit 14.6 Relative value comparison using leverage ratios and credit scores

Relative Value Comparison Using Leverage Ratios

Company Name	Description	Maturity	Price	Yield	Spread	Leverage	Spread/Leverage
Ketchup Co.	6.125% Senior Notes	2021	101	5.95%	356	3.1x	114.8
Mustard Corp.	8.25% Subordinated Notes	2020	101	8.30%	691	4.9x	141.0
Salsa, Inc.	9.125% Senior Notes	2019	103	8.21%	695	6.0x	115.8
Mayonnaise Co.	7% Senior Secured Notes	2020	101	6.80%	465	4.7x	100.0
Soy Sauce Inc.	7.25% Senior Notes	2020	102	6.77%	504	4.7x	107.2

Relative Value Comparison Using Credit Scores

Company Name	Description	Maturity	Price	Yield	Spread	Credit Score	Spread/Credit Score
Ketchup Co.	6.125% Senior Notes	2021	101	5.95%	356	25	14.2
Mustard Corp.	8.25% Subordinated Notes	2020	101	8.30%	691	40	17.3
Salsa, Inc.	9.125% Senior Notes	2019	103	8.21%	695	65	10.7
Mayonnaise Co.	7% Senior Secured Notes	2020	101	6.80%	465	50	9.3
Soy Sauce Inc.	7.25% Senior Notes	2020	102	6.77%	504	55	9.2

The spread per point of leverage or spread divided by the credit score (in which lower scores are better) gives a sense of the implied return of a bond relative to its credit risk. The credit score can be more subjective than the leverage ratio measurement but is more inclusive of the overall credit quality of the credit issuer and the bond.

In the corporate debt market, I have most commonly seen a regression of the spread for each debt instrument as the dependent variable and the leverage ratio as the independent ratio (though a credit score could easily be used, too). A multiple regression could easily be used, too, where several independent variables are factored in to try to determine their influence on the dependent variable (i.e., the spread). Though multiple regression is generally viewed as more complete, it is not as widely used in this type of analysis and it does raise some more potential problems such as auto correlation. However, multiple regressions would certainly be more realistic as there are so many factors that truly go into valuing a security.

Exhibit 14.7 shows some of the data that would be generated from a basic regression analysis in Microsoft Excel. The data was a single-variable regression using 11 bonds and using their spread as the dependent variable and the corresponding leverage ratio as the independent variable.

Exhibit 14.7 Regression statistics of spread and leverage for an 11-bond universe

SUMMARY OUTPUT	
Regression Statistics	
Multiple R	0.923
R square	0.852
Adjusted R square	0.835
Standard error	35.669
Observations	11

The statistics show that the R statistics are strong, over 0.6, which shows that there is a strong predictive relationship between spread and leverage ratios for these issues.

RESIDUAL OUTPUT

Observation	Predicted Spread	Residual
1	363	36.83
2	466	-16.34
3	289	20.52
4	385	-85.28
5	452	28.39
6	297	-1.85
7	481	13.92
8	452	28.39
9	511	-10.56
10	371	-14.54
11	289	0.52

This data gives the predicted spread that the bond should be trading at and the residual shows the difference from the current spread it is trading at to the predicted spread.

Price Movements

The way that markets move the prices of loans and stocks is not always right. If a headline on a company comes out, just because prices move in response to the news does not make that reaction the correct one. However, that does not mean that as an investor you should ignore what the market is doing to the prices of debt securities.

Monitoring the movements in the prices of debt issues can be a valuable signal of danger or can create an opportunity for buying or selling and any systems you set up should watch for meaningful price changes.

Any database system that is regularly pulling in a field for security prices should be able to track the issues you are interested in following and alert you on trades and meaningful price movements in bond and loans. A typical type of alert would be similar to the types that some brokerage services offer for personal accounts, where the system will alert you if there is a price move more than a certain percentage amount, perhaps if a bond moves more than 5% in price in a given day.

However, because of the trading illiquidity in many names in the corporate debt markets, I would suggest a few adaptations to this typical type of alert. Price movements should be tracked over a longer time period than just daily; you should consider monthly or year-to-date or even since a position was taken. You might also want to have absolute prices as an alert, for example 70 or 110.

It is also very important that if you see a single price movement that appears to be an outlier, you check the volume of bonds actually traded and how long it has been since the last trade. Sometimes, a very small amount of bonds can trade and print a price in an illiquid issue, but the trade might not be meaningful enough to react to because such a small amount of debt traded. For example, $100,000 or $250,000 trade in a $500,000,000 issue; this size trade does not make it very meaningful. In addition, if a bond or loan has not traded in several weeks, what may appear as a sudden one-day movement might really just be the pricing catching up with where the market has moved since its last trade.

When you do see an alert about price movement, up or down, in a debt instrument that you are monitoring, it doesn't mean you are supposed to immediately react—it typically is a sign that your team needs to reevaluate or reconfirm your views on the name. You have to first analyze if the relative value changed meaningfully for you to change your opinion and second, you need to analyze if the market is starting to indicate something that you might have missed in your analysis and the overall opinion on the credit needs to be reevaluated.

Using Equity Data

Many of the companies that issue corporate debt have public equity outstanding. Analyzing some of the data on the underlying equities of these companies can be very helpful in analyzing asset valuations and judging the market tone and reactions to headlines as well as a company's access to funding. This is all valuable in analyzing the relative value between different debt issues.

Part of the reason for monitoring the underlying equities is that in almost all cases the public equity shares of a company will have much more trading liquidity than the debt. Additionally, the data on the trading prices and trading volumes on the equities is much more readily accessible than those on the debt instruments.

Monitoring the movements in the underlying equity prices and volume levels can often be a key indicator about market sentiment, either on an industry sector or a specific issuer. Similar to monitoring the price and volume movements on the debt instruments described previously, whenever possible the systems should be set up to monitor the same for equities on both an individual and industry sector aggregate and send alerts for certain movements.

You should also keep in mind that a struggling equity price and/or a rising one might pressure management into undertaking actions that can be good or bad for bondholders. For example, a struggling stock price might prompt a company to look into asset spin-offs, a sale of the company, or other actions, whereas a rising stock price might give a company confidence to issue more stock to deleverage or use its stock, instead of additional debt, to pursue acquisitions.

The equity markets essentially value shares of stock based on what the investors think the company is worth. If a company is sold or liquidated, the equity shares of the company do not receive any proceeds until the debt is paid off first, so that the equity value of a company can represent the theoretical asset coverage for the debt.

Typically, you can add up the number of common shares outstanding on a company and multiply it by the share price for an equity market valuation. If you add in the value of any preferred shares and the net debt outstanding (net debt typically subtracts cash on hand), you have the enterprise valuation for the entire company (EV). As discussed earlier, this is easily compared with the leverage ratio, debt/EBITDA, to get a sense of perceived asset protection.

Remember, EBITDA is effectively a measure of cash flow, and both the EV/EBITDA ratio and the leverage ratio are based off of cash flow. A company may have other value aside from the cash flow it is producing; this is a level of detailed work that a credit analyst needs to study that might not be included in database ratios, but it should be factored into the stock price. For example, a casino company may have two hotel casinos in Atlantic City, New Jersey, that are producing EBITDA of $200 million, but it might also have a large parcel of valuable real estate in Las Vegas and an undeveloped gaming license in Macau. So if the EV/EBITDA ratio is 10×, the equity market is probably not saying the cash flow in Atlantic City is worth 10×, or $2 billion; it is probably saying that the cash flow and the two undeveloped assets together are worth $2 billion. Therefore, you could not look at this multiple and apply a 10× multiple to every other casino company in Atlantic City as a valuation; it would probably be overstating its value.

In a similar vein, sometimes a company may be underperforming with its cash flows, and investors may view it as a takeover target because they believe a different management team could improve the cash-flow margins and, thus, the EBITDA, and this may cause it to trade at a higher multiple, too. There is also the theory that there is a premium paid to control an asset or a corporation (proven time and time again in acquisitions of public companies). This implies that if a company is an attractive takeover candidate for another company that the market multiple may be higher than ones that are not.

A word of caution: Many third-party data sources may provide enterprise value multiples, but you might not agree with how they are calculated. Many treat cash on hand and cash equivalents differently; I have seen some that have ignored preferred shares in their calculation and sometimes miscalculate the right number of shares outstanding (in the money warrants and options as well as separate classes of common stock, which can often lead to meaningfully different valuations).

I would also make it a point to monitor the stock prices of industry leaders in key groups even if they are not companies in which you would invest.

Maintenance Covenants

Tracking and charting maintenance covenants and their headroom can be a very valuable analytics tool, particularly during economic downturns. Maintenance covenants are typical for bank agreements. If the test is violated, there is typically a period to "cure" this "technical default" or the banks and the company may negotiate a permanent or temporary change to the covenant, usually for a fee or a higher interest rate. Being alert to upcoming potential maintenance covenants can be very valuable. This is an area that I believe will become more and more automated and analyzed during the next credit downturn.

There is a certain level of complexity in doing this, however, as each covenant is different, so there are no exact or standard formats to use. Additionally, many loan agreements are not publicly available, and if they are publicly available, they do not all have maintenance tests written in the same way. For example, one bank agreement might have a maintenance test on minimum level of EBITDA and a maximum total debt leverage, another bank agreement might have a minimum interest coverage test and a maximum Senior Secured leverage test, whereas a third bank agreement may have all four of these. However, industries do tend to be somewhat comparable, and when they can be compared, it can be very valuable. Obviously, the proper modeling of the covenant is key.

A typical maintenance test covenant grid might look like Exhibit 14.8.

Exhibit 14.8 Ilustration of a maintenance test and headroom

Maintenance Covenants and Headroom				
	Senior Secured Leverage Ratio			
	Year 1	**Year 2**	**Year 3**	**Year 4**
Maintenance test	4.50x	4.25x	4.00x	3.75x
Ratio as calculated	3.50x	3.50x	3.65x	
Headroom	28.57%	21.43%	9.59%	
	Minimum EBITDA in $MMs			
	Year 1	**Year 2**	**Year 3**	**Year 4**
Maintenance test	500	505	510	520
EBITDA as calculated	625	630	580	
Headroom	20.00%	19.84%	12.07%	

After year 3, the headroom in this example has declined significantly and if the trends continue, a violation could occur in year 4. An alert system would alert you to this, rolling in each quarter's results as they become available.

Once you set up a formula to calculate the ratio that is outlined in the document, you then can calculate the headroom cushion. The headroom is the amount of room you would have for the ratio to move before violating the covenant. Ideally, an alert would arise if the headroom gets under, for example, 10% and there has been a negative trend for two quarters and under 5% no matter what.

Analytics and Nonfinancial Information

What has hopefully become apparent in prior sections of this book is that there are items that are critical to the final analytical conclusion that are not mathematical and are much more difficult to use in analytics. Alerts can be set up to take notice of some of these items, but they are much harder to use in quantifiable analytics. Some factors can be related to the overall debt issuing credit; these might include management history, competitive environment, and perhaps geographic locations. Some credit scoring systems try to capture these. Other factors might include the covenants of the specific debt instrument.

Covenants are not typically issuer specific but issue specific. The quality of the covenants for an issue is very difficult to quantify. We discussed maintenance covenants, but many covenants are "incurrence" covenants where a test or target has to be met before the company can undertake a transaction, such as issuing debt or paying a dividend. Scores could be assigned on inclusion of a covenant (e.g., a point for having a debt incurrence test or a restricted payments test), but this will hardly capture anything about the quality of the language in the covenants and how tough or easy the tests in the covenant are to circumvent.

Some organizations have developed covenant scores that they make available to subscribers, but scores for covenants end up being very subjective and although the scoring attempts are interesting, I have yet to see a method that can help analyze or accurately score how covenants can protect an investor or be circumvented in specific situational cases. For example, how does a leverage test and restricted payments test prevent a company from pursuing a leveraged buyout?

There are other similar items to consider. Perhaps a score could be added for the ownership structure, such as a higher score for a publicly owned company versus a private company or one that was part of a leveraged buyout. Management history is always a major factor as is the competitive environment. Similarly, some basics of corporate structure can easily be scored (e.g., holding company debt versus operating company debt); however, other aspects, such as where individual assets reside, are not as easy.

Many firms do have their analysts add some of these highly subjective points to a scoring system, which is perfectly fine in my opinion as long as all the people using the system understand how and where these subjective points are used.

Endnotes

1. Net free cash flow or NFCF in this usage is EBITDA net of cash interest expense, taxes, working capital, and capital expenditures.

2. Enterprise value in this usage uses the total number of shares outstanding multiplied by the price and then any debt and preferred shares are added to the amount and cash is netted out.

3. Headroom is the amount by which a debt covenant exceeds the test for that covenant. For example, if there is a maintenance test that requires a minimum EBITDA of $100 million and the company has $120 million, it has $20 million of headroom, or 20% headroom. Maintenance covenants are typically structured to require seeing some improvement in the financial ratio for the first few years after the issuance of the loan.

4. Held loan positions are portions of a bank loan that the bank that is the lead arranger (or in the syndication group) either is required to, or just ends up, holding on its balance sheets. Often a requirement in bank loan agreements is for a minimum amount to be held by the bank that is placing the loans; this also helps to assure other investors that due diligence was properly carried out on the transaction. In somewhat perverse regulatory logic, in bond offerings the underwriters are "not allowed" to hold any of the securities they are issuing—the logic being if they are allowed to hold them, they might market time their sales to the detriment of other investors.

Closing Comments on Section V

These chapters covered some of the basic methods in which analytics are used in the market. We started from the top down to derive performance and other trends between various asset classes and analyzed why these differences can impact you. The analysis uses comparison of returns as well as volatility and yields and spreads. Many of these same items are then included in the analysis of a single market; however, the layers by which you can analyze this data are much greater and the amount of data you might want is greater, too. The macro analytics should help your team develop some themes and strategies as well as help you monitor trends in the markets.

Analysts have to keep in mind that when you look at this data, it is often done in time series and the constituents in the market subsectors can change over time. These changes may be the driving impact on the results of a study. Additionally, keep in mind, like so much of statistical analysis, the data you are using is backward looking and the trends you have seen might not be the ones you will see.

The next layer of analytics will help you in credit selection to meet the strategies that you have determined you want to pursue. The quality of this work will depend heavily on the depth and quality of the information that you have available in your databases. This credit selection analysis can not only be valuable in finding issues that meet your criteria, but can also be very valuable in finding outliers that may present risk or opportunity.

Relative value analytics among potential credits to invest in is typically based on a proxy for return on the investment (yield or spread) and credit risk, often measured by a credit score or critical financial metrics. This differs from the relative value as measured by historical returns and the volatility of those returns over time; that is typically used for comparing asset classes or portfolios.

Remember there are major differences between evolving a list of potential investments using a flexible and detailed database and actual credit issue selection. In the description of liquidity analysis, we outlined the difference between the type of data that would be included in a database for analytics and the level of analysis a credit analyst would do to determine if a name on the list would work as an actual investment. Keep in mind that this example only outlined one topic of the many that would need to be explored by the credit analyst.

Section VI
Analysis of Market Technicals

By Robert S. Kricheff

When people talk about *market technicals* for the corporate debt market, they are basically talking about supply and demand in the market. For example, is there significant new issuance coming to market in the form of new bonds or loans and is there significant new buying or selling interest coming from investors as money is attracted to a specific asset class or is it leaving an asset class? Technicals matter greatly in influencing the pricing of securities and overall market movements. They can impact all of the capital markets or can influence subsets of the market—even specific industries, rating categories, or security types can be impacted in isolation from the rest of the market. Supply and demand can reshape a market's profile over time as well.

Broker-dealer trading desks, capital market professionals, portfolio managers, corporate issuers of debt, and asset allocators all want to have an understanding of market technicals to improve their overall performance. A simple example may be that an asset manager has seen some minor redemptions in one of his or her investment-grade funds and then notices from public data that the aggregate amount of assets under management (aum) in investment-grade funds declined slightly over the last two periods. At the same time, a number of investment-grade companies have announced they are increasing their debt issuance. The asset manager might decide that this is a near-term trend and buying new debt now or even holding it will not help his or her performance; the asset manager might decide to raise some cash by selling assets or might at least decide to avoid investing money in the new debt issuance that is coming, as he or she believes even if the new debt is attractive, it will be cheaper to buy in the future because of how the supply-and-demand dynamic is shifting.

Technicals are often influenced by macro events, such as bad economic data or an oil shock, but often the impacts are more subtle. Like a crowded trade within a market, some relationship has flashed on investors' screens as having moved too far in one direction and suddenly there may be a withdrawal of funds from, say, emerging market debt markets. Although technicals usually tend to impact the markets over shorter-term cycles, longer-term technicals can be influenced by many factors as well, such as emerging industries, changing demographics, evolving tax laws, or overall changes in the risk profile of markets.

15

Market Demand Technicals

Introduction

Generally, when trying to monitor the development of technicals, you will be interested in trying to follow several layers on the demand side, depending on your role. You might start at the most macro level of demand and try to see shifts in the general demand for investment in securities, as opposed to bank accounts or another cash proxy such as money market funds. The next layer might be to discern the relative demand for fixed income versus equities. Within fixed income, it is quite likely to try to determine the demand for corporate debt instruments versus other fixed-income assets. The next logical step would be to look at the relative demand among corporate debt products—such as emerging markets, high-yield loans or bonds, investment-grade loans or bonds, or distressed assets—or whether the interest level is to enter the market through structured vehicles such as ETFs or CLOs. Finally, within a given subsector of corporate debt, there might be specific demand trends for putting risk on or not (often colloquially referred to as *risk-on* or *risk-off* trades). The next layer would be demand for certain types of investments or subsectors of the market, such as specific industries or rating categories within these markets.

As you will see, there is not always solid information—or sometimes any information—on the demand for the various sectors outlined previously. However, that does not mean that you will not try to understand and piece together tidbits of data to try to analyze the demand for these sectors while trying to analyze these influences on your markets.

Demand Data

Much of the information you get about the demand side of the equation is backward looking and is limited to only certain market participants. There is, of course, no requirement that every investor needs to say what he or she is doing in the marketplace. However, some investment structures do need to disclose data about inflows and outflows. The problem with the data is that it is far from complete as many market participants do not report this data.

Good detail on the demand side is considerably trickier than on the supply side. Regular data is available for the ETFs and from the closed-end and open-end funds about inflows and outflows; these are available from a number of sources. In addition, significant information about insurance company holdings is available quarterly. In the United States, data has more recently been available about *dealer inventory*, or the size of bond and loan positions that are owned by the large investment banks. However, not all of these funds that report flows and demands may be fully dedicated to debt or one subset of the market (e.g., a high-yield loan fund or emerging market debt fund), so it is unclear where the shifts in the money flows are exactly impacting demand.

Of course, numerous other market participants do not do any public reporting of data on flows or assets under management whatsoever. In addition, many market participants who have assets under management are not dedicated to investing in the corporate debt markets, but might from time to time enter the markets opportunistically. This type of money is often referred to as tactical money. These types of funds can include hedge funds, or balanced funds; also large equity funds often can hold a portion of their assets in corporate debt and general bond funds can often switch between various fixed-income assets. You will often see analysts trying to determine the amount of money invested in an asset class that is strategic versus tactical, meaning which money is dedicated to the asset class long term versus being temporarily opportunistic.

CLO activity can be a critical part of the demand side for loans and needs to be tracked to try to get a handle on technicals for this specific market, though CLOs often invest in secured bonds as well. Public disclosures are not always available about CLO fundings, but announcements are usually made about such activity because it involves marketing of the debt tranches. CLOs (and related collateralized debt obligations [CDOs] and collateralized bond obligations [CBOs]) create demand for debt and the debt tends to be held for long time periods. The increased demand for loans caused by a CLO is relatively short-lived as the CLOs try to fund the collateral over a relatively short time period. However, it is also true that usually when you see one or two CLOs coming to market, others will follow because it indicates that the market conditions appear good for this vehicle. The demand for paper from CLOs in recent years has been a very major factor in the leveraged bank loan market.

Other Demand Impacts

On the demand side, you should not forget the reinvestment effect on the fixed-income debt markets, especially in the higher-yielding markets. If you look at even just one decent-sized, high-yield fund—for example, if there were a fund of $10 billion in size—and assume an average 8% coupon and even a relatively low 50% reinvestment rate, that translates to about $400 million a year in new demand. Keep in mind that the corporate debt market is a multitrillion dollar market. So this simple example shows that the markets can generate an awfully healthy level of demand even without dramatic inflows.

There is also the deleveraging effect. Many companies that have debt outstanding, especially in the leveraged markets, hope to deleverage over time by paying down debt. This may come from equity issuance or from net free cash flow that has been generated from the business or from asset sales. Either way, as this debt is paid down, it produces cash for investors that needs to be reinvested, thus creating more demand for new loans and bonds. This type of demand typically increases during cycles of a healthy economy and/or rising stock market prices.

There is also no question that heavily covered themes and stories in the media can influence demand and contraction for certain asset classes, though more so in equities than in fixed income in my opinion. The articles may scream that the small-cap equity markets are too expensive or that yields on high-yield investments are too low. The problem is the quality of their sources varies, and on any given week you can almost certainly find conflicting advice. This can temporarily influence the market technicals—it does not mean that the conventional wisdom is right, and usually over the longer term it is not, but if you want to do the best you can at your job, these trends need to be watched and analyzed.

Although neither supply nor demand technicals are easy or precise to monitor, I believe that based on the current available information sources, demand is the much harder side of the equation to get details on.

16

Market Supply Technicals

Introduction

The information on supply is a bit easier to monitor and organize than the demand data. This data, however, is still far from perfect. Corporations and underwriters may disclose financing plans in certain cases, but they are cautious about announcements and in some cases are restricted by rules about what they can or cannot say in advance of an actual announcement.

When a new financing (supply) is announced, it might come to market very quickly (the same day) or might take a week or longer of marketing. Loans tend to have a longer process than bonds from marketing to pricing to closing, in part due to the lack of central clearinghouses and because of the amount of documentation that is required.

Most of the information about new supply comes from the underwriters of the supply, mainly investment banks. Any data that your credit research analysts may glean from other news articles regarding perhaps a debt-funded acquisition or a debt-funded expansion could be added to supply information as well. However, any number of things could delay or cancel some of this supply.

New supply is underwritten and syndicated primarily by investment banks; this can include bonds and bank loans. The investment banks tend to publish and keep track of a forward calendar or "backlog" of potential new supply of bonds and loans. These lists generally include their own underwritings and those of others; there are also newsletters that publish similar lists. Some financings, for various regulatory reasons, usually cannot be announced too far ahead of the actual transaction; also sometimes new financing might come to market very quickly, or potential new supply might get "pulled" for lack of interest or a change at the issuing corporation and then the supply does not come to market at all. All this leads to the fact that the forward calendar is far from precise and changes with great rapidity and fluidity.

Use of Proceeds and Other Ways to Analyze Supply

There are various ways to break down the forward calendar that can be helpful in analyzing the impact of the potential new supply. One of the most common aspects is to break down the supply by what the use of proceeds will be. This can help to determine if this forward supply "has" to come to market or does not.

Many times, new issues on the forward calendar can be price sensitive and if the market pricing moves to be less attractive, the supply might not ever come to market. This type of financing is usually for refinancing of existing debt. You should note that in these cases where the new "supply" is a refinancing, it is often creating its own demand because the new supply is being used to take out or retire existing debt, so the retirement of the existing debt creates more demand available for the market. There are times when a refinancing "must" come to market, most notably when there is an upcoming maturity. Some refinancing will be undertaken because the overall interest rate environment has become more attractive and the company can save money by issuing new debt to take out old debt—this type is most market sensitive. Refinancing because the credit profile has changed dramatically (or perhaps because a higher-quality company now owns the debt-issuing entity) is less market sensitive.

When the debt-issuance markets are strong, companies might also look to issue debt to return capital to their equity holders. This can typically come in the form of a dividend or in the form of stock buybacks. When supply is coming to market for these types of purposes, the issuance is only modestly price sensitive and if the pricing is far off of expectations, these types of deals may get pulled and disappear from supply.

Typically, if the proposed financing is new capital for an announced acquisition, by a corporation or a financial leveraged buyout (LBO), or an announced project, you can usually assume that the financing has to come to market in some way and during some time period. Obviously, if problems crop up and the acquisition falls apart before closing, the supply disappears. Similarly, if markets change dramatically, a project might be abandoned and the financing might not come to market.

When a company is looking to raise money for general corporate purposes, it is somewhat harder to generalize how market sensitive the issuance is. It is a bit of a catchall designation and a financing with this designation for the use of proceeds might be simply to give the company added liquidity, to take advantage of attractive financing terms, or for more strategic reasons. Therefore, these types of funding need to be explored more closely to see if they should be considered a "must" come to market or not.

You can analyze the potential supply of new issuance on the forward calendar in many other ways. Often, new issuance comes in waves in a particular industry. This supply pressure can reprice an entire industry sector of a market. Potential supply can also be analyzed by risk categories, notably security ranking and rating. When a particularly large amount of supply is expected by ranking or rating, it does not usually put pricing pressure on those segments of the market the same way supply in an industry might, but it does give insight into the tolerance for risk in the market.

Often, the different technicals of two markets can influence supply. Most commonly, I have seen this between the high-yield bond and leveraged loan market and the U.S. dollar and the euro markets. When the demand, and pricing, gets stronger on the loan side, you often see a planned offering of debt-substitute loans for bonds and the reverse when the opposite is true. There are trade-offs of course; loans have a floating-rate structure, often have more covenants, and tend to be shorter maturities. This same substitution effect happens often when a European company is issuing in both U.S. dollars and euros. Depending on the demand, the amount of new financing might shift from the originally proposed amount more toward one currency or another. Usually, the European-based company has a bias toward issuing in euros to balance its assets and liabilities and also because there is usually the added cost of hedging the U.S. dollar currency debt.

Exhibit 16.1 shows how new issue data might be laid out so that it can be analyzed. Note in particular the two columns on the far right, which show the related debt retirements that are expected and are likely to create natural buying demand for the new financings.

Exhibit 16.1 An example of a forward supply table

Illustration of a New Issue Supply Summary

(Issuance Amount in Millions)							Related Retirements	
Company	**Description**	**Maturity**	**Bonds**	**Loans**	**Timing**	**Use of Proceeds**	**Bonds**	**Loans**
Lions, Inc.	Senior Secured	7 year	500	0	This month	Acquisition	0	0
	Senior	8 year	300					
Tigers Corp.	Senior	10 year	500	300	This week	Refinancing	500	300
Bears, Inc.	Senior	8 year	500	400	This week	LBO	300	200
	Senior	10 year	300					
	Subordinated	10 year	200					
Leopard Corp.	Senior	8 year	200	100	This week	GCP	0	100
Platypus Co.	Senior	7 year	400	300	Next 2 weeks	GCP	0	0
Total Supply			2,900	1,100			800	600
Net Amount			2,100	500				

GCP=General Corporate Purposes
LBO=Leveraged Buyout

Analyzing Price Talk and Pricing

Monitoring other aspects of new issues can also give you other insights into supply and demand and, specifically, the types of issuance that may come to market. Notably, monitoring the pricing, sizing, and the post-issue trading can all give indications of the overall strength or weakness in a market and what types of financings buyers are looking for.

When a new issue comes to market, there is usually price talk. The *price talk* is an indication of where the issuing company and the underwriters expect to bring the new financing. The indication is usually given in yield or spread. For example, suppose a new company that makes pallets is coming to market with a $400 million high-yield bond deal. The underwriters may come out with price talk of 7.5% to 8%, indicating that would be the yield to maturity on the notes when priced, or they might give the price talk in spread of 550bps–600bps. However, spread is more often used on investment-grade deals or when the bonds would be trading much tighter to the non-risk benchmark (e.g., treasury note) so the spread would be tighter. It is worth noting a few things about the price talk: When given in yield, this often equates to what the coupon will be on the new bond; however, sometimes the bond will get issued with a lower coupon. The bond will be issued at a discount, for example, a 7.25% coupon, but instead of coming at par (100% of face value), it will be sold at a price of 99.25% of face value to equal a yield of 7.5%. Also price talk is not typically put out in an announcement. The underwriters, sales force, and capital markets teams disseminate the information to potential clients, so it is not usually readily available from third-party sources unless a news service, like Bloomberg or Reuters, picks it up. There is sometimes early price talk just as the new issue is announced, and then there is official price talk closer to when the financing is actually going to be priced. Often in the offering prospectus, the pro forma financial data may show an expected interest rate on the new financing, which is where the early whispers of price talk often come from, or it might be based on the yield levels of other bonds that the company has outstanding.

It is valuable to track and analyze price talk versus the actual pricing. In a strong market, the yields on the new bond will likely be at the low end of the price talk or might even be lower (also referred to as tighter) than the price talk (for example, in the preceding example, say the new issues come at 7% at par). In a weak or a weakening market, it might come at the wide end or even wider than the talk. Obviously, when tracking this data, it is important to be consistent in comparing the final pricing with official price talk, and the data needs to be kept in the same format (i.e., yield or spread).

Any one issue can have certain characteristics or nuances that can cause pricing to vary from the price talk to the actual closing, and some underwriters are better at gauging the right price talk than others. So, when you design a system to monitor this data, the analysis should use a rolling 10- or 20-deal history; you might want to eliminate especially small deals, for example under $200 million or €150 million in size, as these might not give real indications of market demand.

Although typically less volatile than pricing levels, changes in deal size are another indication of the demand for that particular type of deal and for the overall asset class. Sometimes the actual amount of the bonds for sale will change prior to the final pricing. It could get increased or decreased depending on the demand.

Other factors can be negotiated by the buyers and the underwriters that can impact the demand but do not get reflected in a change in price. Most notably, covenants might get changed or a call structure might get shifted. This can improve the terms of the bonds for the buyer meaningfully, but will not reflect itself in a price change. For example, using the previous case, perhaps the deal was struggling to get priced at 8%, but the underwriter and the issuer did not want to raise the rate and were able to convince buyers that 8% was a fair price if they moved the first call date back from a standard five-year, non-call period to a seven-year, non-call period and raised the basic debt incurrence test from a 2× interest coverage test[1] to a 2.5× interest coverage test. These changes would not be observable in monitoring the price talk and pricing, but clearly show that the market was somewhat resistant to the new issue.

Other structural items can give you some color on where there is the greatest demand—notably, maturity dates. In high yield, for many years it was common to see 10-year maturities, but now depending on the issuer and the market demands, 5-, 7-, 8-, and 10-year maturities are all relatively common. In the investment-grade debt markets, the range is even wider and 20- and 30-year deals are regularly completed. The loan markets tend to have shorter tenors. A trend toward longer-dated issues typically indicates a strong market and more comfort that interest rates are either stable or possibly expected to go down.

The size of the forward calendar itself can be an indicator of market strength as well. Although it is not always the case, a growing forward calendar is usually an indication of a strong market. When new financings are pulled and do not come to market and the calendar shrinks in size, it is usually an indication of a weakening market.

People should also be aware of the cycles to new issuance as well. Not surprisingly, the forward calendar and new issuance tend to be weaker during periods when many people take vacations such as around August and December, and then typically pop in size after those periods. There are also quiet periods during the year when companies are in the process of finishing their final results for a period and could not release them in an offering document.

Postplacement Trading

Another very important indicator of the market strength and receptivity to new supply can be how new issues trade after they are priced and distributed in the market. If the new issue is a bond, most times the underwriter and the issuer would like to see the bond trade up slightly, perhaps a percentage point or two. If it trades down or trades up meaningfully more than that amount, it can be an indication of the strength or weakness of the market, or of how well the specific transaction is priced relative to demand. So again, I find it best to track this on a rolling average, perhaps the most recent 10 or 20 deals.

I think the timing of when you examine the pricing is also important; taking the price on the day it is issued or one day after a deal prices is good to look at, but there is often a fair amount of noise around the early trading of a new issue. Tracking the price a week or two weeks after it was initially priced might give a better indication of where the real level on the bond has settled in. Given the limited liquidity in the market for many issues, it is not uncommon to see money managers track these postissuance trading levels by underwriters as well because the lead underwriter historically provides much of the initial liquidity. Investors can then rank their performance.

Investors sometimes take particular note of how certain specific large-sized issues or "benchmark" names may trade. For example, if there is a headline-making LBO that comes to market with a $5 billion term loan B and a $3 billion Senior Unsecured Note, investors often look at how well this transaction is received and how it trades as a bit of a bellwether for the market.

Supply and demand can also be impacted by debt retirements or ratings upgrades and downgrades. As mentioned earlier, new financings are often used to take out existing debt; sometimes this substitutes existing supply in one market, for example the high-yield bond market, for new supply in another market, for example, the leveraged loan market. Upgrades and downgrades from the rating agencies can also shift supply from specific markets. For example, most people consider a bond with a BBB- and a BAA rating from Moody's and S&P to be investment-grade bonds and anything rated below that to be high yield. So if a bond is rated BA/BB and through a series of events gets upgraded to "triple BBB," theoretically this is supply of debt leaving the high-yield market and going into the investment-grade market. However, these types of supply shifts are usually much more gradual than those from new issuance or debt retirements. Most investors are able to own a portion of their portfolio in assets out of their asset class, at least for a period of time. Additionally, the rating agencies usually move in increments when upgrading issues so that a large BA/BB-rated high-yield bond may gradually trade into investment-grade funds over a year or more before it is upgraded to actual investment grade. There are exceptions, however. Event-driven ratings changes can cause the supply to shift hands more quickly. Some examples of a sudden event-driven ratings change could be a company with debt outstanding that is rated single B suddenly being bought by a strong investment-grade company, or perhaps an investment-grade company announcing a large, leveraged special dividend recapitalization or leveraged buyout. In these cases, the supply tends to shift significantly more rapidly usually accompanied by significant price movements in the debt instruments.

Supply and Demand Impact the Face of the Market

All the changes to supply and demand in these various markets impact the analytical work that you will want to do. Over relatively short periods of time, it can change the overall makeup of the marketplace. This can impact how you want to manage your positions in your portfolio and how and where you want to allocate your analytical resources.

I believe it is important to keep track of how supply and demand is shifting your market tiers, too. Keep in mind that throughout the capital markets, there is often a pack mentality: If a new loan structure is successful, investment bankers will increasingly pitch that to their clients; if an industry that previously had not accessed the high-yield bond market has one or two companies successfully issue debt, others will look to do so as well. Therefore, during a very active new issue market, the overall makeup of the market can change dramatically.

There can be waves of new structures that can shift the market. The structures can include many features. Often, the newer features appear first in the high-yield loan and bond markets, such as equity clawbacks, pay-in-kind bonds, toggle bonds, and so on. Similarly, in the leveraged loan markets, we have seen the evolution of various call protections and covenant-lite issues.

There have also been time periods where certain rankings enter the market more, perhaps a rash of Senior Secured Notes that are being used to replace loans as opposed to the Senior Unsecured Notes that had dominated the markets overall. Sometimes the most obvious of the changes are within industry groups.

All of these types of changes can change the risk structure of the market over time and should be monitored and analyzed over a longer time period (semiannually or annually would be my suggestion). These types of changes can increase the volatility or duration of the overall market and you will need to reevaluate how you position your own investments in a given market relative to the overall changes in the market.

Endnote

1. An interest coverage test is a typical covenant test that needs to be met to be able to issue more debt or perhaps undertake another similar type of transaction. A typical interest coverage test would measure the ratio of EBITDA/total interest expense.

Closing Comments on Section VI

The technicals of a market can have a major impact on the performance of trading positions and portfolios over shorter-term cycles. It can influence the decisions of when to buy and sell. Even though the information is far from perfect and monitoring it can be a bit like piecing together a patchwork quilt, these technicals are very important.

Market participants need to be able to monitor the supply and demand as best they can and should regularly get and examine data analysis on the supply and demand. This can help immensely in making buying and selling decisions. I think the role and importance of supply and demand have increased dramatically. It is not surprising that just a few years ago, it was unheard of for money managers or private equity firms to have a dedicated capital markets person, and now it is more and more the norm. It is an important weapon in your quiver and should not be ignored. Longer term, the changes that can be wrought on the market by technicals can impact performance even more. You will be at an advantage if you monitor the direction the market is going and how its profile is changing ahead of time.

There are investors who are primarily market timers and look to play market technicals, often through CDS indexes or, more recently, through ETFs, and sometimes through certain no-load funds or money managers. There have certainly been cycles when there has been high correlation among the vast majority of credits in a market and where dispersion of spreads and yields are tighter than historical averages. During these cycles, it often seems that following technicals and the top-down approach to investing is all that is needed; however, in the longer run, credit quality still is the key. If a credit cannot service its debt, no amount of short-term market strength will help it over the long run. Although I do not favor a market-timing strategy as a dominant investment philosophy for long-term investors, I also believe that all market participants need to monitor and react to technicals as part of their decision process whether they are short-term or long-term investors if they want to perform the best they can. To this end, remember much of the data that influences market technical analysis is spotty; part of the key to analyzing this information is to have it well organized in a readable, structured fashion, have it analyzed regularly as it is very fluid, and, finally, try the best you can to not make decisions about these factors based on hearsay or gut instinct.

Even if your group is only focused on supply and demand for one part of the corporate debt markets, you should monitor supply and demand for similar markets. These similar markets can often be substitutes from which supply and demand can shift back and forth.

Section VII

Special Vehicles—Liquid Bond Indexes, Credit Default Swaps, Indexes, and Exchange-Traded Funds

By Miranda Chen

Since the beginning of the twenty-first century, several important structured vehicles and derivative products related to the corporate debt markets have seen a boom in demand and assets dedicated to them. It is critical that you understand these products so that you can include them in any analytics that you undertake. Liquid bond indexes may have led the way for some of these products—and from them, corporate debt exchange-traded funds and credit default swap indexes have evolved, which together are some of the most liquid instruments in the corporate debt markets. Because of their liquidity, they supply some of the most rapid and real-time color on trends in the market. However, they are not the complete market; they are just a portion of the market. They have also increasingly become hedging and arbitrage vehicles, so both their structures and their own innate technicals create certain idiosyncrasies in how they trade and act. If you understand some of these aspects, you can better determine how to use them in your analytics.

17

Liquid Bond Indexes

Introduction

As should have been apparent from Section IV, "Indexes," creating a corporate bond index to gauge the broad movements of the credit markets is difficult. Even when compared with the U.S. equity markets where 3,000 securities in the Russell 3000 cover over 97% of U.S. listed companies by market capitalization, to cover 95% of the U.S. corporate bond market, you would have to track over 10,000 investment-grade bonds and 2,500 high-yield bonds. Even more challenging, money managers and investors need to account for the fact that on any given day, only 400 corporate bonds trade with more than a $10 million notional size. This, of course, does not even begin to address the leveraged loan market. This creates a quandary for investors in terms of benchmarking performance because of the lag in prices and the failure to account for transaction costs for less-liquid bonds.

Modern corporate bond indexes are relatively young when compared with equity indexes. Lehman Brothers created the first total return–based corporate bond index in 1973, which calculated the coupon and price return of corporate bonds on a monthly basis. However, it was not until 1986 that a daily bond index was created.

Why a Liquid Bond Index and What Is It?

The general problem with corporate bond indexes is there is no way an investor can actually own this portfolio of bonds. The liquidity of the majority of the constituents in the index is quite low and it would be difficult to go into the market and buy all 4,000+ investment-grade or 2,000+ high-yield bonds. Investors also generally prefer highly liquid bonds, not securities that are difficult to buy or sell (for instance, trade once a week). To meet the demand of most investors, sell-side security firms created investable liquid bond indexes in the early 2000s. The goal of these bond indexes was to reflect the tradable (liquid) portion of the corporate bond market. However, because it's difficult to know ex-ante what bonds will be liquid, index creators must use objective criteria beforehand, which usually consist of (1) size and (2) years from issuance restrictions.

For instance, two widely quoted liquid bond indexes—the Barclays (formerly Lehman) Liquid HY Bond Index and the Markit iBoxx Liquid US High Yield Bond Index—require bonds that are included to be issued within the past five years and be at least $500 million or $600 million in size. The purpose of the size minimum is due to the generally greater liquidity of large versus small issues, whereas the years from issuance filter accounts for the fact that bonds that are not as recently issued may become "locked away," meaning investors are essentially holding the bond to maturity in the traditional sense of locking the bond in a safe until it matures.

What Are Benefits and Drawbacks of a Liquid Bond Index Versus a Full Index?

The benefit of using a liquid bond index over a traditional full index is that the additional restrictions—such as size, years from issue date, and issuer capitalization—create a narrower set of bonds, which better reflects the tradable and investable market.

Many fixed-income portfolio managers are benchmarked against a public fixed-income index (e.g., the Barclays Aggregate Bond Index). These investors seek to match their portfolios to the benchmark by purchasing and holding many of the cash bonds in the benchmark. Clearly, if an investor were to build a portfolio from scratch, he or she would have a higher likelihood of sourcing the bonds for the liquid corporate bond index in a short amount of time and would find the liquid index easier to replicate and rebalance as new bonds are issued and added to the index and existing bonds drop out due to ratings-based criteria. The fact that the bonds are larger and more recently issued means that they are more likely to be traded and have smaller transaction costs.

Importantly, a number of exchange-traded funds (ETFs), marketed by Blackrock iShares and State Street Global Advisors, which are discussed in more detail in Chapter 19, "Corporate Debt Exchange-Traded Funds (ETFs)," reference these liquid bond indexes, thereby enhancing the trading frequency of the constituent bonds in the index and making this subset of the market the most liquid in the corporate bond universe.

A whole area of analytics has risen around these tradable indexes and ETFs. Quantitatively driven trading firms and sell-side banks try to predict the additions and deletions to the liquid bond indexes because these ETFs are benchmarked to the liquid bond indexes and will likely follow with an equivalent buy and sell once the bond is added or removed from the liquid bond index. Sometimes, the buy-or-sell order is expected to be quite significant and causes the bond to trade on its own technicals driven by this ETF-driven supply and demand for a short while. For instance, most liquid bond indexes have a certain minimum size for the bond to be eligible for the index, so if a new issue bond is above this criterion (e.g., $500 million), then it will be added to the index and ETFs will very likely buy the bond. Thus as an investor, it is important to know the rules for the major corporate

ETFs so that you can be aware of the trading dynamics of a particular bond based on its inclusion in the liquid corporate bond index.

One key disadvantage of using the liquid bond indexes is the shorter time history of the liquid benchmarks. For instance, the Barclays High Yield Index traces its roots to 1986; however, the Barclays Liquid High Yield Index only began in 2002. Thus, the benchmark you choose to use can depend on the age of your portfolio. If you have a portfolio that goes back greater than ten years and want to use liquid indexes as your benchmark, you need to switch to the broader, nonliquid benchmark for periods prior to 2002.

For reference, Exhibit 17.1 compares some of the rules and criteria of a typical U.S. High Yield Index with a typical Liquid U.S. High Yield Index. Although the liquid bond index is able to track the broader index with a high correlation and minimal tracking error, far fewer bonds are needed to build this portfolio.

Exhibit 17.1 Comparison of liquid versus broader U.S. High Yield Index

Benchmark Index	U.S. High Yield Index	U.S. High Yield Very Liquid Bond Index
Index rules		
Issuer cap	No limit or a fairly high limit	Two or three of the largest bonds per issuer
Size	Bond size >$100 million	Bond size >$500 million
	$150 million or $250 million are typical	Or greater
Maturity	> 1 year until final maturity	> 1 year until final maturity and usually must be issued recently within the last 5 years
Rating	Typically uses a middles rating of Moody's, S&P, and Fitch or average rating	Similar rules

Benchmark Index	U.S. High Yield Index	U.S. High Yield Very Liquid Bond Index
Excluded	Noncorporate bonds, structured notes, bonds with equity-type features (converts, contingent capital securities), Eurobonds, defaulted bonds, private placements, floating-rate issues, emerging market bonds, non-U.S. dollar bonds	Similar rules
Included	Fixed-rate puttable and callable corporate bonds, 144A securities, zero coupons, PIKs, fixed to floating perpetual bonds, fixed to floating structures, while in their fixed bond period	Similar rules
Defaults	Usually excluded	Similar rules

You might argue that because the broader index and more restrictive liquid index have very low tracking error, what need is there for a liquid benchmark? If you are an investor who holds larger bonds and has difficulty sourcing smaller deals, you will find that over time your performance will lag that of the broader, less-liquid index. Based on our analysis, we have found that smaller deals (<$300 million) have outperformed, as these issuers need to compensate investors for their additional risks; that is, liquidity premium. Additionally, given the lower level of trading volume, these smaller issues are likely not to be priced as accurately, which causes less volatility in the index during certain cycles, even though the lower volatility is not based on actual trading. Therefore, as a money manager, your risk-adjusted performance will often lag a benchmark.

18

Credit Default Swaps and Indexes

What Other Tools Do Investors Use to Measure the Corporate Bond Market?

Corporate bond market participants have struggled with the issue of liquidity for some time and the decline of broker-dealer balance sheets since the 2008 financial crisis has not helped. Outside of cash bond indexes, investors, money managers, asset allocators, and traders look at other tools to monitor broad movements in the corporate bond market:

- ***CDS Indexes.*** In particular, the CDX Investment Grade and CDX High Yield Index
- ***ETFs.*** In particular, the iShares and State Street Global Advisors investment-grade and high-yield indexes

As you can probably tell by this part of the book, the reality is that the majority of the corporate bond and loan market does not trade every day, so market participants need to find other tools to quickly assess how the broad movements of the market have affected the price of their holdings and portfolio (the beta). Some market participants have argued that because these CDX and ETF instruments are traded with greater frequency, these indexes reflect the transactionable price in the market at any given time. Because of the limited liquidity in the corporate bond market, many market players view trading with ETFs and CDX as the only way to quickly adjust the risk of your portfolio. The traded volumes on the CDX HY index for instance can be 3× to 5× the actual cash market (Exhibit 18.1), making the CDX index one of the fastest ways for a large institutional manager to swiftly add or pare exposure to the markets. In fact, many institutional investors use these vehicles to rebalance their positions and risks, depending on their investing restrictions, even though this is removing them at least one step from investing directly in the corporate debt.

Exhibit 18.1 Comparison of HY liquidity

	ETF	Cash HY Volume	CDX HY Volume
Daily volume	$300–500 million	$5.0–$7.0 billion	$20–$30 billion
% of HY cash market size	0.05%	0.50%	1.5%

What Is CDS and What Is a CDS Index?

CDS, or credit default swaps, were designed to protect lenders from losses in the event of default on the part of the borrower by transferring risk to another entity in exchange for a series of regular payments. One way CDS has traditionally been viewed is as *bond insurance,* in which the buyer of the CDS is purchasing an insurance policy from a trusted counterparty (CDS seller) that would protect him from losses should the referenced issuer of the bond default on the bond by either missing a bond coupon payment or filing for bankruptcy. The buyer of the CDS makes periodic payments (usually quarterly) during the contract period (usually five years or less) and in the event of a default, the buyer of the CDS receives a payment from the seller equal to the notional size of the contract but delivers the underlying security to the seller of protection. *Importantly, buying protection on a CDS contract or on a CDS index means selling risk, and selling protection means going long risk.* Exhibit 18.2 illustrates the required payments for a CDS buyer and the contingent payments for a CDS seller.

Exhibit 18.2 CDS mechanics and payments

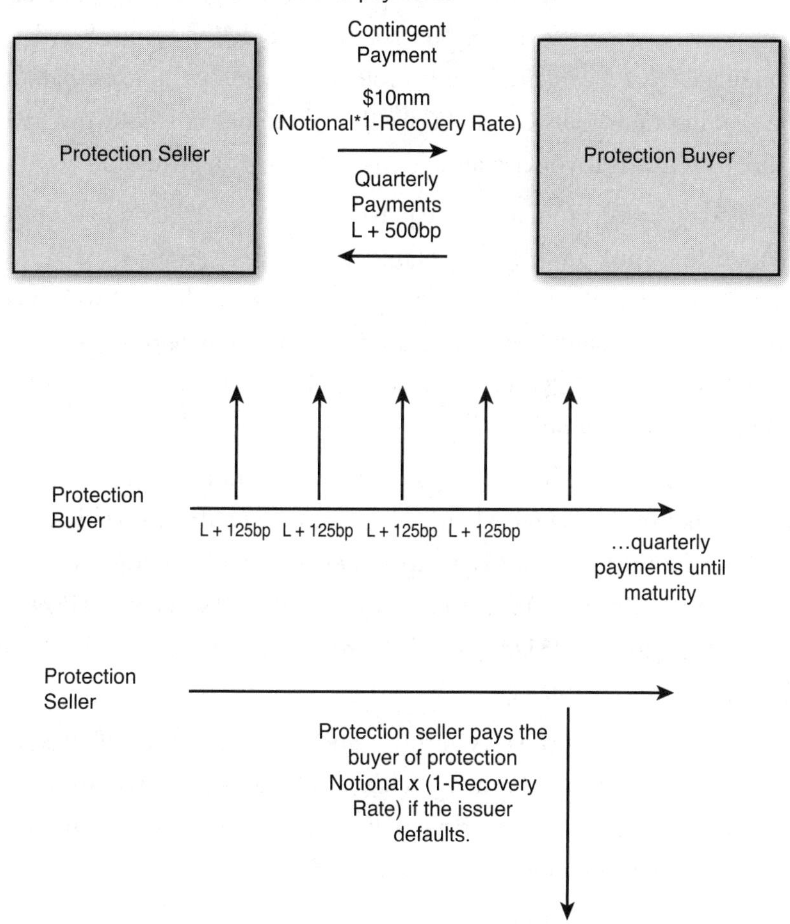

Similar to a liquid bond index, you can buy contracts that represent an index of some of the most liquid CDS contracts. The most popular index of contracts traded in the United States are CDX (credit default indexes) while in Europe the iTraxx indexes are the most prevalent. There is a CDX index for North American investment grade CDS, CDX IG, and one for high yield, CDX HY, that are widely followed and traded. In Europe, the investment grade index is known as the iTraxx Europe and the high-yield series is known as the iTraxx Crossover, which contains HY issuers. The market convention is for indexes that trade on a spread term (e.g., CDX IG, iTraxx Europe, iTraxx Crossover) to be quoted in protection terms. So when you buy CDX IG, you are buying protection, whereas indexes that are quoted in price terms CDX HY are thought about in risk terms, so if you buy CDX HY, you are adding to your long-risk position.

Today, however, many more participants want to buy default protection than just lenders and banks selling protection. A variety of market participants, such as hedge funds, money managers, insurance companies, and banks, can freely use CDS to speculate or hedge individual company (single-name) risk as well as portfolio risk with the CDX IG and HY indexes.

CDS trades are over-the-counter (OTC) transactions, meaning they do not trade on an exchange, and before a contract is traded, the two parties must enter into a swap agreement that has been standardized by the International Swaps and Derivatives Association (ISDA), referred to as the ISDA Master Agreement, which sets the legal framework for trading.

Over the years, CDS contracts and conventions have become standardized, and standardization has accelerated with increased regulation since the financial crisis in 2008. Each CDS contract has the following specifications:

- ***The Reference Entity.*** This is the underlying company or entity that the buyer/seller of protection is transferring risk on.
- ***The Reference Obligation.*** This is the security (bond or loan) of the reference entity that the buyer would like protection on. Although this bond is not the only deliverable bond of the entity, it represents the lowest seniority bond that is deliverable.

- ***Maturity/Tenor.*** This states the term of the CDS contract, which is most often five years, although most dealers quote any tenor up to ten years.
- ***Notional Principal.*** This is the size of the contract/size of risk referenced.
- ***Credit Event.*** This states the criteria for triggering the CDS contract. In the United States, the CDS contracts trade with *hard default* language, meaning the reference entity missed a coupon on its obligations or filed for bankruptcy. However, in Europe, the CDS trades with *soft default* language, so the CDS could be triggered if the company simply restructures its debt.
- ***Coupon.*** This refers to the annual premium paid to buy/sell protection. This is quoted in basis points (bps) running and is generally 100bps for investment-grade companies and 500bps for high-yield companies.

Usually, it is quite obvious when a CDS event is triggered, but a CDS participant must officially bring the credit event before an ISDA vote to start the CDS auction and settlement process. So, if you have purchased protection on an issuer and a CDS event is triggered, you usually must wait a few weeks before you are paid.

Understanding CDS Pricing—Spreads Versus Prices

CDS contracts for nondistressed companies trade on a spread basis. This represents the effective annual fee to buy or sell protection. The higher the spread on a CDS is, the more risky investors believe the company is and the higher its likelihood of defaulting. The spread is calculated by taking the default probability and multiplying it by the default loss rate of the bond. For instance, a CDS spread of 100bps for a one-year contract would imply the credit has a 1% default loss probability (1.67% chance of defaulting × 40% recovery rate), and a 300bps spread would imply a 5% change of defaulting, assuming the same 40% recovery rate. Of course, the probability of defaulting is not equal each year; usually the probability increases as time goes by, so CDS curves are usually upward sloping. Although actual coupon payments are fixed, up-front payments are made on the contracts to adjust for the risk differences. A common screen that CDS traders use to confirm these up-front costs is the Bloomberg Professional CDSW screen.

An investment-grade CDS generally trades with a 100bps running coupon, whereas a high-yield CDS traditionally trades 500bps running, although both IG and HY CDS will be quoted in spread terms. However for some distressed issuers, CDS traders will quote the contract in up-front prices. This assumes a running contract of 500bps and is the up-front cost you pay to enter the contract (in the CDSW screen, it is in the lower-left box).

Setting a pre-agreed-upon recovery level for the CDS is a pivotal part of valuing the contract because it represents the value postdefault and, therefore, impacts expected cash flows. Market convention since 2009 assumes a 40% recovery rate for the trading purposes. In the event of default, the auction process determines the actual final value of the CDS for settlement purposes.

CDS Indexes—How Are They Constructed?

So far, we have discussed the basic mechanics of a single-name CDS. An index CDS is simply a portfolio of single-name CDSs chosen to represent the broader market. When CDS indexes first began trading in the early 2000s, a number of competing firms (sell-side and independent) were vying to be the benchmark index, much the same way the S&P 500 is the primary benchmark for U.S. equities. Through consolidation and attrition, Markit Partners runs the only liquid set of CDS indexes traded today. The Markit CDX North American Investment Grade CDX Index, with 125 constituents, and Markit CDX North American High Yield Index, with 100 constituents, are the industry standard for trading U.S. CDS indexes, while the Markit iTraxx Europe, with 125 constituents, and Markit iTraxx Crossover indexes, with 50 constituents, are the prime corporate CDS indexes in Europe.

The current series of the index trades with a three-, five-, seven-, and ten-year maturity and the indexes *roll* or extend their maturities by six months every March and September 20th in order for the current, or on-the-run, index to keep a constant maturity. (Currently, the CDX HY indexes roll one week later on the 27th.) Each roll date, the constituents of the CDX and iTraxx indexes are rebalanced to account for corporate events, such as mergers, acquisitions, and defaults as well as ratings changes. Issuers that no longer qualify for a particular index are deleted and added to the eligibility of other indexes if appropriate (e.g., an issuer that is upgraded to investment grade is no longer eligible for the CDX HY index and will be removed).

The CDX indexes are designed to include the most liquid single-name CDS for each benchmark by design. When an issue is removed from an index, a replacement is chosen based on liquidity. The Depository Trust & Clearing Corporation (DTCC) aggregates all CDS settlements and publishes a semiannual report before the roll, which lists the most traded, single-name reference obligations by rating group. To avoid high turnover every roll, new entities are usually only added when constituents are removed. Prior to March 2013, the additions and deletions to the roll were conducted via a scheduled dealer poll, but this procedure was modified to be more rules based so that the indexes could trade on a futures exchange.

What Can We Gauge from CDX Pricing and Skew?

The CDX indexes normally trade in line cash markets, when the market is rallying. CDS and cash spreads both move tighter, and conversely, when the market is selling off, spreads in both markets widen. However, CDX indexes only capture default risk and risk premium. Additional risks such as liquidity risk and interest rate risk that are inherent in cash bond investing are not captured. Additionally, the betas of the cash and CDX markets are different, and over time one can significantly outperform/underperform the other. This is illustrated in Exhibit 18.3, which compares the spread of the CDX HY index with a generic HY cash index. As you can see, while the correlation of the CDX indexes to the cash market is high, the CDX indexes tend to move faster and can sometimes move in opposite directions to the cash market due to technicals caused by the buying and selling pressure from the diverse group of credit market participants. At times, CDX can also be more resilient than the cash market due to the fact that CDS only captures default risk and risk premiums. For instance, during the interest-rate-driven sell-off in May and June 2013, both HY cash and CDX HY sold off sharply. However, CDX HY, which is insensitive to interest rates, recovered more rapidly than cash once investors realized the driver for the sell-off was only the fear of rising rates and not a slowdown in the economy, which would drive up default risk and risk premiums.

The difference between the traded price of the index and fair value of the index, the price determined by summing the constituent spreads of the index, is known as the index skew. The skew tells us a great deal about the dynamics in the market place. For instance, when the CDX HY index trades above fair value, the index is said to be "trading rich," and this suggests a strong risk appetite for HY credit at the moment because investors are eager to add risk and willing to pay more than the price as calculated by underlying single-name CDS contracts that make up the index. If, on the other hand, we see that the CDX HY index is "trading cheap," this usually implies that the risk appetite is weak because dealers are willing to sell the index below its fair value. Other market technicals can drive the index to trade rich or cheap to the underlying single-name CDS, such as strong hedging needs from other markets (see "The Maiden Lane Case"), all of which would be captured by the skew, so skew remains an important gauge for market risk sentiment. This metric is comparable to investors monitoring whether ETFs trade above or below their net asset values (NAVs), which is discussed in Chapter 19, "Corporate Debt Exchange-Traded Funds (ETFs)."

Exhibit 18.3 High-yield cash versus CDX spread difference

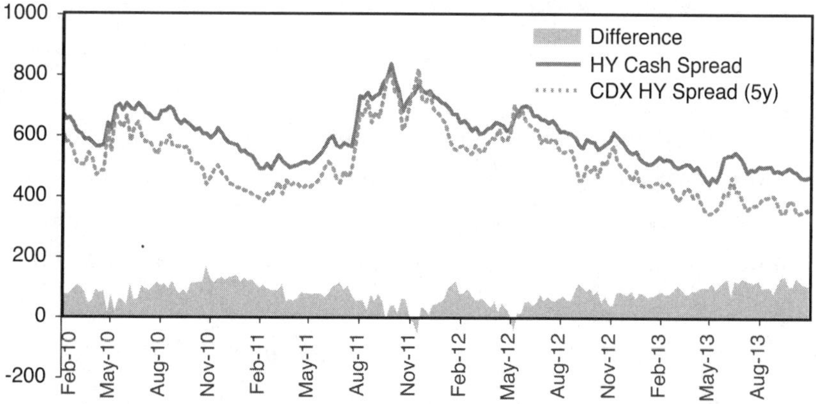

Who Are the Participants?

Since CDS indexes began trading a decade ago, their popularity has increased manyfold. Today, CDX indexes are the fastest way for investors to quickly add or hedge their credit exposure due to their liquidity. This includes superior liquidity and better bid-offer spreads to many other vehicles. More than $70 billion CDX IG and $20 billion CDX HY trade on a weekly basis, solidifying these two instruments' rank as the two most liquid securities in the corporate debt universe. Recent volumes are shown in Exhibit 18.4.

Exhibit 18.4 Comparing the liquidity of cash bonds versus single-name CDS

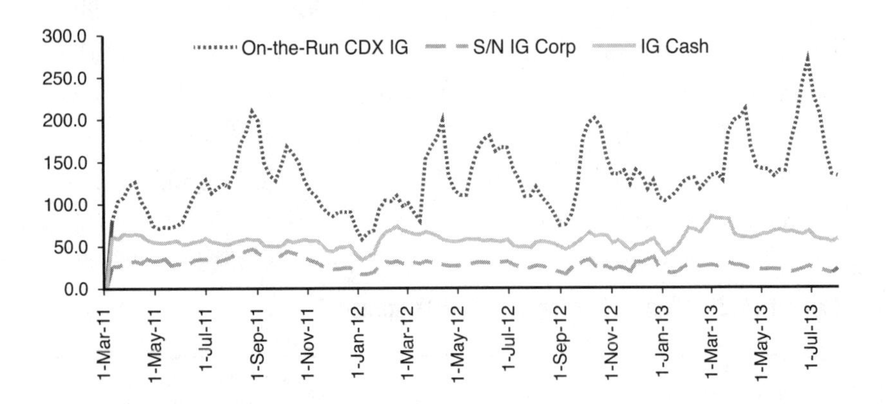

A broad swath of market participants, including corporate bank trading desks, hedge funds, money managers, insurance, and sovereign wealth funds, are all active in the space. However, most mutual funds have restrictions on owning derivatives and cannot use the CDX indexes for hedging and rebalancing purposes.

Corporate bank trading desks are the intermediaries between investors and balance the buy-and-sell needs of their clients, including hedge funds, money managers, and insurance and sovereign wealth funds. Most corporate bank desks do not hold onto large positions overnight, relative to the volume they trade during the day. The Depository Trust & Clearing Corporation (DTCC) publishes trading volumes and net dealer positions for most of the indexes, from which we see that dealers typically trade three to five times the volume they hold in inventory. These reports can be very helpful for analytics on market activity and direction, even if you and your team are not active in CDS or CDS indexes, as they are the most active. Particularly in high yield, it can be argued that the volumes and directions in the contracts represent the more tactical investors in the market who can influence the most recent trading levels in the most liquid issues, as opposed to the longer-term holders of these instruments.

When corporate bank trading desks need to offset their own risk, they go to interdealer brokers. These brokers intermediate trading among large banks (that is, JP Morgan, Goldman Sachs, Bank of America, Barclays, and Credit Suisse).

Investors who use the indexes trade with the banks and use the indexes to reposition their portfolio quickly. For instance, if a money manager receives funds from a new pension mandate of $200 million, the money manager can sell CDX IG (sell protection, long risk) as a placeholder while the manager builds up his portfolio. The manager then gains exposure to a diversified pool of corporates with one $200 million trade with a bid-offer spread of 1/2bps as opposed to trying to buy 100 $2 million positions of bonds that would cost 5bps–10bps in bid-offer spread. Similarly, if the manager needs to reduce risk, perhaps because he or she thinks volatility will pick up over the next month but feels comfortable with his or her bond positions, the manager can hedge the portfolio by buying CDX IG protection and not have to liquidate bond holdings to raise his or her cash position with the 5 bps–10bps transaction fee. Of course, this path of getting the $200 million invested does not work for everyone all the time. The trade-off in this example is that the pension mandate that the money manager was given is not being placed directly in bonds but in an index of derivatives. The pension that allocated the money to the manager likely has many other pools of investments and may have another manager charged with placing directional bets on the investment-grade market, while this $200 million was to be invested more directly into credit picks with a specific profile as part of overall diversification for the pension.

New participants for other areas of fixed income, for instance, counterparty valuation desks (CVA) and mortgage trading desks, have also dipped their toes in the CDX and iTraxx markets for regulatory capital relief and hedging needs. This has, at times, distorted markets because these players can build large positions and are less price sensitive, similar to when an ETF comes in to buy a specific bond.

The Maiden Lane Case

There was an interesting case in May and June 2011 that shows the value of monitoring the CDX products and carefully analyzing the available data on them. During this period, the Federal Reserve was starting to unwind the portfolio of risky mortgages that they had assumed from American International Group (AIG) during the credit crisis to bail out the deeply distressed firm. These portfolios, known as Maiden Lane II, after the street that runs alongside the New York Federal Reserve, contained nongovernment guaranteed residential mortgage-backed securities (RMBS) totaling $20 billion. As Wall Street banks' mortgage trading desks rushed to hedge the technical headwind associated with this large unwind by the Fed, they bought CDS protection in their market, namely with CMBX and ABX, but ultimately moved into CDX HY given the greater liquidity in the CDX HY index. Ultimately, this caused the index to underperform in May and June and had repercussions on the HY cash markets as investors wondered if the CDX market was foreshadowing another slowdown in the economy.

Implications and Limitations of CDX

More than $70 billion in CDX IG and $20 billion in CDX HY trades on a weekly basis, making these securities the largest and most liquid instruments within the IG and HY credit universe, respectively. Because they are traded so frequently, even traders and investors in the corporate bond markets who do not actively trade these instruments monitor their intraday movements. The skew of the index can also help investors gauge risk appetite. As these indexes trade more frequently than cash bonds, they are great overlay tools for adding and reducing risk, and they are also incredibly useful for spotting a turn in market sentiment. Often, when at a pivotal point in the market, you will see the indexes change direction before the rest of the market.

As discussed with various players in the CDX markets, fast money participants, such as hedge funds focused on indexes, have developed algorithms to take advantage of intraday volatility and positioning. This can cause intraday fluctuations that are not explainable by treasury or S&P moves. Additionally, as the scope of participants in the indexes is expanded beyond U.S. credit investors, to mortgage or CVA participants, their hedging needs can differ from that of traditional market participants and cause disruptions, what market participants refer to as "technicals." The Maiden Lane example earlier demonstrates how CDX HY weakness was triggered by mortgage hedging and did not correspond with activity in our market. Thus, while the correlation of the CDX indexes to the cash market is high, the CDX indexes tend to move faster and can sometimes move asynchronously with the cash market due to technicals caused by the intraday buying and selling pressure from the diverse participants in the credit market.

19

Corporate Debt Exchange-Traded Funds (ETFs)

What Are ETFs?

ETFs are investment vehicles that combine the diversification benefits of mutual funds, intraday exchange liquidity of stocks, transparency of costs and holdings, and tax efficiency. Equity ETFs have been very popular for decades as ETFs like SPY, QQQ, and IWM have become part of mainstream media with hundreds of millions of shares traded daily. Fixed-income ETFs were introduced in 2002 and have thrived in recent years as assets ballooned from $20 billion in 2007 to over $200 billion in 2013. Credit ETFs represent the largest share of growth with IG and HY ETFs accounting for ~40% of the assets in fixed-income ETFs.

One way of thinking of ETFs is as a bond asset that gives instant diversification at a fraction of the cost of trading the same basket of bonds. Credit ETFs were revolutionary when they were first introduced in 2002 because this was the first time retail investors could see live pricing for a diversified and tradable basket of investment-grade and high-yield bonds. Before credit ETFs were introduced, investors could buy and sell cash bonds, which have a face value of $1,000 as the minimum trade size. However, because of this minimum trade size, large for retail but small for an institutional market, mom-and-pop investors paid much higher bid-offer costs and typically invested in mutual funds to get corporate bond exposure. With the first credit ETF, the iShares iBoxx$ Investment Grade Corporate Bond ETF (Ticker: LQD), and the first HY ETF, the iShares iBoxx$ High Yield Corporate Bond ETF (Ticker: HYG), retail investors could transact for cents on each share, effectively gaining access to the market in real time for a fraction of the cost.

A second way to think of an ETF is as a mutual fund that is closely benchmarked to an uninvestable index, with intraday trading. For instance, the ETF LQD is benchmarked to the Markit iBoxx USD Liquid Investment Grade Index, and the ETF HYG is benchmarked to the Markit iBoxx USD Liquid High Yield Index. Unlike a mutual fund, which only provides end-of-day pricing and requires you to place a purchase or redemption order before market close, and without knowing the exact price that you will pay, ETFs trade on an equity exchange and provide full transparency on the bid and ask price, usually less than $0.05.

In summary, the advantages that ETFs offer that have attracted so much growth are as follows:

- ***Intraday Liquidity.*** ETFs provide access to intraday liquidity through the secondary market, which is not available for index mutual funds (which only trade based on end-of-day pricing). Closed-end funds exist but are much smaller with very limited liquidity.
- ***Cost Efficiency.*** Purchasing an ETF is a cost-effective way to gain exposure in the credit markets because secondary market bid-offer spreads for ETFs are in pennies. This contrasts with 1/4 to 1/2 point spreads ($0.25–$0.50) in the underlying bonds. ETFs can have lower expense ratios than the lowest-cost index mutual fund. For example, the Barclays iShares $ Investment Grade Corporate Bond ETF (LQD) has a management fee of

15bps, whereas the Fidelity Intermediate Bond Fund (FTHRX) has a management fee of 32bps.

- ***Instant Diversification.*** Current IG and HY ETFs hold a basket of 500 to 1,000 bonds, lending to instant portfolio diversification.
- ***Hedging Capabilities.*** Selling or shorting ETFs is becoming an effective way to hedge portfolio risk because CDX might not be as representative with only 100 or 125 reference names.
- ***Administrative Simplicity.*** Investors can purchase a diversified pool of bonds on a listed exchange with one electronic transaction rather than hundreds of transactions with the same pool of bonds.

ETFs also have several drawbacks. There tends to be little to no active management, so there tends to not be credit analysis underlying the bond selection, as selection tends to just follow inclusion rules. When market swings are volatile, the ETFs can feel the brunt of the moves as they tend to trade in the same direction as the market, which can look good at times or very bad at times. Additionally, the portfolio does not typically look at or care about concentrations in types of securities, so you could be buying into a portfolio with limited diversification.

Mechanics of ETF Construction

For credit ETFs to take in or return cash, you need to go through an authorized participant (AP), which is usually an investment bank, and deliver or receive a sub-basket of bonds that the ETF is looking to buy or sell that day. This method of creating or redeeming ETF shares is known as in-kind delivery, as opposed to cash delivery where the ETF shares are exchanged directly for cash. This might not appeal to all investors, especially smaller investors who would have trouble sourcing the underlying bonds. The creation and redemption process should not be confused with the secondary trading of the ETFs. Investors can still buy and sell the ETF on listed exchanges where there is usually ample liquidity.

Exhibit 19.1 details the fund's creation/redemption mechanics with all the participants involved. The main thing you should note is that the investor is responsible for buying/selling the underlying bonds that the ETF manager uses for creating/redeeming shares.

Exhibit 19.1 Credit ETF mechanics

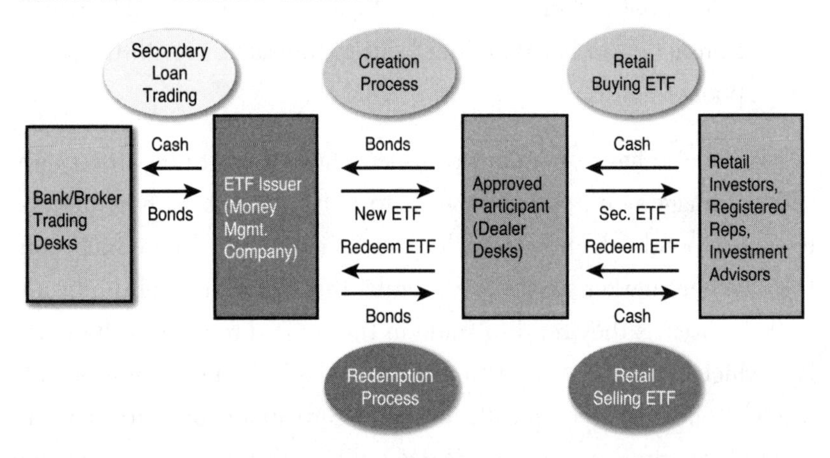

The benefit for the ETF provider in doing this is the ETF provider will always be able to meet investors' creation and redemption needs because the investor is responsible for buying and selling this subset of bonds. This is also beneficial for secondary market participants because it forces the ETF to trade relatively in-line with its Net Asset Value (NAV). For instance, if the ETF ever trades at a significant premium to the NAV, the authorized participant can buy the sub-basket of bonds for in-kind creation and receive the ETF and gain from the difference in the process.

Secondary trading in IG and HY ETFs is similar to trading equities or equity ETFs. They are listed on an exchange and trade during normal trading hours with a bid and offer and indicative size these quotes are available for. For larger transactions, you would contact the ETF trading desk of the investment bank directly.

ETFs Extend into Leveraged Loans

One area of credit that has experienced explosive ETF growth in 2013 has been leveraged loan ETFs. Leveraged loans have gathered significant interest as investors looked to take shelter from rising rates and associated volatility. Loan funds and loan ETFs have been a major beneficiary of this trend. The PowerShares Senior Loan Portfolio (Ticker: BKLN) is one of the best-selling ETFs in 2013 and has accumulated $5.0 billion of assets with over $3.5 billion in 2013 alone. However, a number of newcomers, including SPDR Blackstone/GSO Senior Loan (SRLN) and the Highland iBoxx Senior Loan ETF (SNLN), have garnered interest. Bank loan ETFs provide a few unique features that differ from other credit ETFs that have been described.

One major difference between the loan ETFs and HY ETFs is the creation and redemption process, which is done via cash rather than delivery of shares (in-kind). This places the responsibility of liquidating loans on the fund manager. Because loans are not securities, they are not as easily transferrable. Secondary sales are conducted via assignments or participations with investors usually trading through dealer desks at the large underwriting banks. Assignments typically require the consent of the borrower and agent, whereas in a participation agreement, the buyer takes a participating interest in the selling lender's commitment and the lender remains the official holder of the loan. You can imagine how cumbersome it would be to transfer 130 $1 million tranches of loans, let alone the assignment fees, which range from $500 to $3,500 per trade, involved.

In addition, traditionally loan investors can choose to receive private-side information, confidential material that banks and accredited investors receive about managers' projections and estimates during the syndication process that public-side investors are not privy to. Retail investors would not be qualified to receive this information, hence making the in-kind transfer impossible. So the fact that this cash creation and redemption process is used, in which the manager takes in funds and purchases loans directly from bank loan trading desks, makes administrative and financial sense.

Due to these differences in creation/redemption mechanics, the ETF issuer passes along the cost of buying and selling loans to the approved participant (AP). As NAV is marked at business close based on mid-market prices, the ETF administrator typically purchases the loans the next day and passes along the mid-offer costs. The fee can range depending on market liquidity and execution costs in BKLN. SRLN has a cap of 15bps for creations and 25bps for redemptions. For larger transactions, the cost can be considerably lower than 25bps and offers a cost-effective way for investors to quickly add/decrease their exposure to leveraged loans.

ETFs in Asset Allocation

ETFs have an increasingly important role in asset allocation schemes as they allow asset allocators to rapidly ramp up or reduce their exposure to credit in a diversified and representative way. This is especially true for private wealth managers' funds with less than $15 or $20 million to dedicate to credit exposure. Pension and institutional-sized funds with greater than $25 or $50 million could probably create separately managed accounts directly with fixed-income asset managers for lower fees; although asset management companies are increasingly offering options for these investors to go into specialized funds that resemble a specialized managed account.

However, a midsized asset allocator might have difficulty replicating the diversification that the ETF offers from scratch especially as the liquidity in the market has declined and the top 100 bonds account for the vast majority of trading. Therefore, gathering a portfolio of 600 HY bonds or 1,200 IG bonds would take weeks or months. Aside from the administrative burden and need to set up trading relationships with Wall Street trading desks, the size of the positions might be too small for institutional trading desks and be considered odd-lot. Odd-lot trades incur larger bid-offer spreads because the size of the trades are less than what is considered standard (<$1million). Of course, history has shown varying drawbacks in performance with an ETF versus other options and less ability to direct any investment rules.

ETFs Used as a Measure of the Market—What about Technicals?

ETFs can and often do diverge in price from their NAV for various technical reasons. The NAV of an ETF is calculated by dividing the total value of all the securities in its portfolio by the number of its shares outstanding. The market price of the ETF, by contrast, is determined by the forces of supply and demand. An ETF is said to be trading at a premium when its market price is higher than its NAV; simply stated, you are paying a bit more for the ETF than its holdings are actually worth. This implies that the market for the underlying securities is in strong demand and accounts for the costs of reconstructing the portfolio, which includes the transaction and administrative cost. By contrast, when an ETF is trading at a discount to its NAV, you' re buying the ETF for less than the value of its holdings. When these conditions arise, the market is usually declining and dealers are less willing to hold onto the ETFs for fear that the liquidation value of the underlying bonds in the ETF is less than NAV should they redeem directly with the ETF manager.

Closing Comments on Section VII

It is hard to imagine doing any detailed type of analysis on macro trends in the corporate debt markets without understanding CDS and CDS index products as well as ETFs. They are two booming areas of the corporate debt markets. They also both offer some of the best detailed and timely data on pricing and trends on the largest and most liquid issues in the markets. Data on these products should be regularly tracked and run versus a portfolio or trading position on both leading and lagging basis for correlations of performance. Also do not ignore tracking volumes in these products versus performance in the broader market as well to get a better sense of liquidity. Ideally, trends should be posted daily.

The products offer significant liquidity for investors, but we also believe they offer certain risks and difficulties. Increasingly, they appear to not just be straight investment vehicles, but also hedging and arbitrage tools for institutional hedge funds and others. The assets that are piling in to these vehicles appear to come much more from technical and tactical investors rather than from strategic and fundamental investors. At times, it is likely their use by these investors may dominate how these vehicles act. There also will likely be cycles where these vehicles' lack of credit selection will impact them significantly.

However, with all the attractions of these instruments and the caveats of the preceding paragraph (and many that have not been mentioned), it is very important that these vehicles not be ignored as a measure of intraday trading and short-term movements in the market. Whether the people you work with invest in these assets or steer away from them, you must include them in any data analytics that you complete on the markets.

Section VIII
Collateralized Loan Obligations (CLOs)

By Robert S. Kricheff and Alexander L. Chan

Collateralized loan obligations are a huge portion of the bank loan market. They are structured and asset-backed vehicles that buy loans. Specifically, they tend to buy leveraged loans. CLOs depend on being able to use market and portfolio analytics to be able to construct and monitor their portfolios. CLOs typically require a disciplined and conservative approach to be successful.

Although CLO issuance can go through cycles, depending on the spreads between various assets in the market, generally they are a growing asset class. With the increase in the number of CLOs, there will likely be an increase in the amount of secondary trading in the debt that CLOs issue (also known as CLO tranches) and the CLO equities. This will also likely warrant an increase in the analytics used to measure and gauge the creditworthiness and return profiles of CLOs.

20

Introduction to CLOs

What Is a CLO?

A collateralized loan obligation (CLO) is a stand-alone entity (usually a special-purpose vehicle [SPV]) that owns a basket of corporate leveraged loans. The loans are the assets of the CLO. To fund the purchase of the assets, the CLO issues debt to investors. The debt makes up the liabilities of the CLO. The loans are the collateral that backs up the value of the debt tranches that the CLO issues.

The debt is typically issued in tranches with the senior-most tranche being rated AAA because of the excess value of the assets over this liability. Although the size of the senior-most tranche has varied over time, as a percentage of the total capitalization, a range of 60%–70% of the total capitalization of the CLO is not uncommon in recent times. The junior-most tranche is effectively the equity and receives any residual value after all of the other tranches are properly treated and paid. The basic idea is not significantly different from many other asset-backed/securitized financing vehicles, such as mortgage-backed securities (MBSs).

Why Do They Matter for You?

The process of structuring a CLO involves heavy use of good data analysis. Models have to be built that value the collateral of the CLO and can track the changes in the values of these assets. These models also have to be able to model the cash-flow streams of the assets relative to their ability to service the liabilities.

These streams of cash flows and the collateral need to be stress tested against certain default scenarios; in other words, assume a certain portion of the assets default and then factor in their recovery rates. Within these scenarios, you need to observe how well each debt tranche is covered and also run various return scenarios for the equities.

You also need to construct the portfolios so that on a blended basis, the portfolios meet certain maturity constraints and diversification rules. Because the debt issued by the CLO has certain maintenance covenant tests, you must set up a system for how well these tests are being met; all transactions are monitored by third-party trustees.

As you can see, data analytics is heavily used in the process of structuring and running a CLO, and if you work in a data analytics group in corporate debt, run loan portfolios, or sell or trade loans, you do need to understand this product. Additionally, although it is not terribly liquid, there is a growing secondary market for trading the CLO's tranches of debt. Trading the CLO tranches requires similar analytics. Managers, investors, investment banks, buyers, and sellers in the secondary market; lawyers; and trustees involved in CLOs all either have to have these types of analytical models, or at the very least understand the models of others in the field.

If you are an investor in the loan markets, even if you do not get involved in CLOs, you need to understand the market for CLOs. CLOs are one of the major sources for demand in the leveraged loan market for second-lien tranches of debt. If you are going to try to analyze and monitor the leveraged loan market and try to understand some of the trading technicals of the loan market, you definitely need to watch what drives the ability to fund these products and analyze the data because it will impact how the market trades.

A variety of software packages run many of the analytics that are needed for the actual administration and covenant management of CLOs. However, virtually every firm also has a certain number of models and systems of its own to manage certain aspects of the process of structuring, buying, selling, and monitoring these instruments.

Some Basics Affecting CLO Issuance

Two of the keys that market participants need to monitor are the spread that you can create between the basket of leveraged loans and what has to be paid out to the tranches of debt that are issued by the CLO. This arbitrage is a key to the ability to issue CLOs. For example, if the spread demanded on AAA-rated CLO tranches widens too much relative to the rate on leveraged loans, it can be very difficult for CLO structures to work. Another key is the level of default loss rates, the loss an investor incurs once a company defaults, but net of the recovery rate. An extended period of very low default rates may lower default assumptions and this could lead to a wellspring of CLO issuance. The reverse is also true, too: A high level of defaults can put an abrupt end to CLO issuance—even if the models might not immediately change, it can frighten investors away. This is because the CLOs are overcollateralized, meaning the underlying assets at face value are worth more than the liabilities, but if there is a higher-than-anticipated number of default losses in the CLO asset pool, this overcollateralization can erode very quickly. If this starts to happen, CLO issuance can come to a shockingly abrupt halt.

Types of CLOs

There are some basic types of CLOs; but once you get past the basics, there are all types of nuances in the structures.

First, there are static and managed CLOs. The static ones maintain the same pool throughout the life of the transaction. The much more common kind is the managed CLO, which has a collateral pool that changes as a manager decides to buy and sell assets during the life of the structure.

There are also balance sheet and arbitrage CLOs. Balance sheet CLOs are usually created when a financial organization, often a bank, wants to get assets off of its balance sheet, for regulatory purposes or other reasons. The pool of assets that is owned by this type of CLO comes from this one source of supply. The arbitrage CLO is more common. The arbitrage CLO is a product typically developed and marketed by money managers to take advantage of the additional income that the portfolio of loans offers over the average interest rate on the debt tranches that the CLO issues.

There are cash-flow and market-value CLOs. The difference between these types of structures is based on how the collateral pool of leveraged loans is monitored relative to the liabilities. In the cash-flow model, the maintenance tests and ratios are measured assuming the assets are at par with a few exceptions like distressed assets. In the market-value CLO, the collateral is marked to market, or based on where the loans held in the pool could likely be sold. Cash-flow CLOs have historically been and continue to be much more common.

Finally, there are synthetic and cash CLOs. Synthetic CLOs hold loan credit default swaps (LCDS) as the assets. Cash CLOs have become the vastly more common type of CLO. These own the actual leveraged loans as opposed to the derivative CDS product.

In the recent post-2008 financial crisis world, CLOs have evolved into primarily being of the following types:

- Managed
- Arbitrage
- Cash flow
- Cash

You should be aware of the other structures used during other market cycles, as you never know what might come back through in different cycles of the market.

Although these are the basic types of CLOs, the nuances within the structure and the variation of the types of covenants are endless. The nuances change as the market changes and depending on who the manager or underwriter is. The demands for different features can also depend on the balance between supply and demand for the product as well.

21

Structure of Typical CLOs

Introduction

As alluded to earlier, the liabilities of each CLO can have numerous major and subtle differences from each other. But a typical CLO will have about five or six tranches of debt and an equity layer. The ratings will range from AAA to an unrated equity tranche and the coupons of each layer of debt issued by the CLO will go up as the ratings go down. Note, with a typical CLO, the coupon on both the tranches of CLO debt and the assets[1] will be floating rate, so movements in rates should not be an issue. Based on the recent cycle of issuance, many of the assets will have a LIBOR floor, so these assets may not "float" until a certain level of LIBOR is reached (or whatever benchmark is being used).

Exhibit 21.1 shows an example of a $400 million transaction based upon late-2013 deals. Please do not think that these ratios or rates have been the same historically or will be the same in the future. Also keep in mind that the debt tranches can be issued with a small original issue discount;[2] more recently, though, if this does occur, it is usually seen in the more junior tranches.

As in any typical debt issuance, the tranches of a CLO may or may not come at a discount (that is, issued below par). If there is a discount, then the actual yield, or discount margin, would be higher than the stated coupon rate. For recent transactions, it is typical for

CLO tranches to be priced with a discount, especially the lower-rated tranches.

Exhibit 21.1 Illustration of CLO structure

Tier of Debt	Coupon Rate	Rating	Amount in $Millions	% of Capital
Class A	LIBOR +145	AAA	250	62.5.%
Class B	LIBOR +220	AA	50	12.5%
Class C	LIBOR +300	A	20	5.0%
Class D	LIBOR +375	BBB	20	5.0%
Class E	LIBOR +500	BB	20	5.0%
Equity	Excess cash flow	Nonrated	40	10.0%

The assets, or the collateral pool, are made up of leveraged loans and are typically not in place when the CLO is initially funded (the CLO is funded when the liability tranches are all placed), but the assets are purchased over time. Typically, the pool will have characteristics such as the following:

- 100–150 issues/loans (sometimes even more)
- Exposure to 20 or more industries
- An average credit rating in the mid-single B-rating categories
- Any principal payments received due to maturity or an early call get reinvested in assets for the first three to five years of the CLO

It sounds fairly simple, but, ultimately, the pool of assets has to pay out more than the CLO's cost of its funding, which is the total cost to pay interest on its liabilities, and the value of the assets has to exceed the value of the liabilities at maturity.

More Analytics: Tests and Measures

Like covenants in a bond or a loan, the CLO structures have strict rules and maintenance style tests that need to be met. Unlike many bond maintenance covenants, when certain tests are not met in a CLO, there are actions that can and often must be taken to rectify the problem. Usually these actions lead to a change in the portfolio composition, including asset sales, until the violation is cured. This composition change can sometimes be to the detriment of more junior tranches.

If problems become too severe, the liabilities are usually structured to be paid out in strict priority ranking and may be paid by delivery of the assets as well. The equity tranche typically has the vote to close a CLO completely and call the debt, but there is usually a non-call period in the structure that encompasses a few years.

The following list looks at some of the typical tests and quality measurements used in analyzing a CLO:

- ***Weighted Average Rating Factor (WARF).*** Originally developed by the rating agency Moody's, this measurement is used to compare the average credit rating of a portfolio. The measurement is weighted for the size of each holding. Also, the points for each rating category do not move up in a simple, straight-line format; for example, the factor for an Aaa rated bond is 1, for an Aa1 is 10, for an Aa2 is 20, and for a Caa3 is 8070. As you can tell, a lower WARF implies a higher credit quality as measured by the major rating agencies.
- ***Diversity Score.*** Also originally developed by Moody's, this score measures the diversification of a portfolio by both industry sector and issuer. Moody's uses approximately 33 industry groups and factors in the correlation of different industries with each and the size of the industry sectors. The higher the score, the more diverse the portfolio is considered.
- ***Overcollateralization Test (O/C).*** As mentioned earlier, the value of the collateral (i.e., the assets) needs to exceed the value of the debt that is issued by the CLO in order to raise financing. The test measures the underlying collateral value at par versus

the tranche of debt in question and all classes senior to the tranche in question. The language of the test usually requires only performing (i.e., nondefaulted) collateral be included in the test, and sometimes if the proportion of very low-rated loans or loans trading at a very large discount exceed a certain percentage, they are not included in the test or included at a haircut in the valuation. If the test that is included in the CLO agreement is not met, certain remedies may be required to be undertaken, such as using excess cash flow to pay off senior tranches until the ratio is met or using the excess cash flow to buy excess assets until it is met.

- ***Interest Coverage Test (I/C).*** This ratio checks that the collateral pool generates enough interest income to service the payments on the debt tranches. Similar to the collateralization test, this is run for each tranche.

These are just a few highlights of some of the tests in CLO structures. There are also typically weighted average spread tests, a test for country of domicile per issuer, and weighted average coupon tests as well as many other tests that get run.

As you can see, good analytics for both the CLO manager and CLO buyers are key in structuring the portfolio, monitoring the portfolio, and selling the CLO liabilities. It is hard to do data analysis on the leveraged loan market without understanding these products.

Endnotes

1. Assets here refers to the loans that the CLO will purchase.
2. When loans or bonds are issued, they are typically sold at par or 100% of face value; if they are sold at a lower price than that—as an example 99% of face value—they are said to be issued at "original issue discount."

Closing Comments on Section VIII

A CLO is an alternative way for an investor to get exposure to the leveraged loan asset class. The investor gets the ability to have various degrees of risk depending on what his or her tolerance is. Compared with investing in a mutual fund of leveraged loans versus a CLO tranche, both give the investor exposure to a very diversified pool of assets with a single investment. In buying a mutual fund of leveraged loans, an investor might get more liquidity to be able to withdraw his or her money, but all investors are *pari passu*[1] with all of the other investors, so there is no way of adjusting to a different level of risk. With the CLO, the investor can choose the level of risk he or she wants and invest in the tranche of debt from the most senior investment-grade-rated piece to the riskiest equity tranche depending on his or her risk tolerance and return goals. However, the CLO tranches tend to be much less liquid, certainly than mutual funds, and less liquid than the underlying collateral.

Despite the relative illiquidity of the CLO tranches, there is an increasingly active market for secondary trading in CLO tranches and equities. As you can imagine, the buyers and sellers of the tranches do considerable analysis on the strength of the collateral and the tenor of the debt tranches. It is important to note that if there does have to be a CLO unwind, it follows strict priority and there is little to no room for bickering among the tranches.

What Does Analyzing CLO Data Tell Us?

There is certainly a large school of analysis that goes into the CLO product, but you can also learn much about what is going on in the leveraged loan market, and to some extent the high-yield bond market, by monitoring and analyzing the amount of issuance and the type of issuance that is happening in the CLO market. When CLOs are very active, they have a material impact on leveraged loan issuance because they can create significant demand for new issuance. They also have a significant tendency to be very limited on price flexibility. If spreads on new leveraged loan issues become too tight, the CLO simply might not buy certain loans and can be very sensitive to the maturity schedule of a new loan, too. Additionally, there is a significant substitution effect between leveraged loans and high-yield bonds. When CLO activity is strong, corporations might shift issuance more toward leveraged loans than bonds, which can lead to an increase in refinancing of existing bonds and a shortage of new supply to replace this product.

These are just some of the factors that you need to weigh when analyzing the data that is available on the CLO market. CLOs are becoming more mainstream, but they remain alternative products with many nuances in structure. Nonetheless, investors are paid multiples of what other similarly rated products yield.

Endnote

1. *Pari passu* is a Latin phrase that means "at an equal pace" or "at an equal step"—in finance it means that two instruments have the same claim. For example, two corporate bonds that are *pari passu* would have an equal claim on a company's assets.

Section IX
Tools for Portfolio Analysis

By Alexander L. Chan

A well-diversified, well-managed portfolio is the best way to achieve consistent returns and growth in assets. Therefore, monitoring the diversification and management style of an investment portfolio, and making adjustments when necessary, is essential to meeting long-term investment objectives.

A thorough portfolio analysis is an important tool for investors and portfolio managers alike to gain insight into the "how and why" of a portfolio's return. Beyond a simple total return comparison, a detailed portfolio analysis allows investors to periodically assess if the investment is meeting or exceeding their investment goals as originally planned in terms of risk tolerance, asset allocation, time horizon, and return hurdles, among other investment objectives. The goal of the portfolio analysis is relatively simple: to understand a manager's performance and whether the decisions he or she is making are appropriate given the investment objectives. However, sell-side trading desks and risk managers can also use many of these tools to analyze the typical holdings that are on their books as well.

22

The Why, What, and How of Portfolio Analysis

Goals of Portfolio Analysis

The goals of a portfolio analysis might vary somewhat from investor to investor or from manager to manager, but ultimately, everyone is seeking to break down and understand the portfolio's return. A comparison is usually made against a benchmark. Therefore, the very first and arguably most important step of portfolio analysis is to determine an appropriate benchmark for the portfolio. Although it might seem like a simple task to choose an index with the same types of securities as in the portfolio, an index usually does not exist that is a perfect match for each manager's style and asset mix. Choosing an appropriate benchmark is critically important as the analysis of the portfolio's risk and return will be greatly influenced by the benchmark.

Historically, investors sought managers that have a style and history that fit their investment criteria. More recently, however, many investors decide on their asset allocations by studying and comparing the returns of broad market indexes first, then choosing a manager in the asset class to which they have chosen to allocate. Therefore, when the investor does finally choose a manager, he or she often expects the manager to perform broadly in-line with the asset class benchmark because that is the return series the investor used to determine the asset allocation. Within the equity asset class, this comparison may be a reasonable approach. For most widely available equity benchmarks, a manager can (and in some cases actually does) buy all the underlying components of the benchmark index. For example, an equity manager can buy all 30 equities in the Dow Jones Industrial Average or even all 500 stocks in the S&P 500 Index. This is possible because equities offer sufficient liquidity to do so. Even the smallest name in the Russell 2000 Index (the 2000 smallest companies of the Russell 3000 Index) trades daily.[1] However, for many fixed-income asset classes, the benchmarks are generally not replicable due to liquidity constraints. For most nongovernment fixed-income asset classes, the number of bonds that trade on a daily basis is significantly smaller than the total number of bonds in the index. For example, there are over 2,000 individual issues in most of the major high-yield indexes, but most high-yield portfolios hold 300–800 issues. For example, one of the largest high-yield mutual funds is Vanguard High Yield Corporate Fund with over $16 billion in total assets, and it holds less than 500 individual high-yield bond issues.[2] This distinction between equities and most fixed-income asset classes is crucially important when thinking about the asset class and appropriate benchmarks.

Even after an asset allocation is determined, careful consideration of a benchmark is still needed. Each index provider determines its own calculation methodology, inclusion rules, and pricing source. Therefore, the performance of indexes within the same asset class can vary greatly over time. This is especially true for credit asset class indexes, such as high-yield bonds. An index provider must answer many criteria. When are new bonds added to the index? How is the index priced? When are rising stars taken out? Does the index include defaulted names? Is there a minimum size? All of these factors can lead to different returns for the same asset class. In addition, managers may specialize in a specific area. These areas of expertise may include sectors like high-quality bonds, distressed and defaulted bonds, a specific rating category, or a specific industry or set of industries. If this is the case, perhaps a sub-index would make a more reasonable benchmark.

As discussed earlier, equities generally provide enough liquidity to actually purchase all of the underlying stocks in your benchmark. However, for fixed-income securities, it can be vastly different. For corporate bonds, including investment-grade, high-yield bonds and loans, only a fraction of the market trades on a daily basis, and those that trade with enough size to create a decent size portfolio are even more limited. In fact, the Financial Industry Regulatory Authority (FINRA) estimates that the daily average trading volume for high-yield corporate bonds was $5.8 billion for the first seven months of 2013. Admittedly, not all trades are captured; for example, many trades of private Rule 144A securities. Even so, given the current estimated market size of the U.S. dollar high-yield market is over $1.25 trillion, it is clear that only a small fraction of the market trades on a daily basis in both volume and number of bonds. The takeaway with respect to portfolio analysis and benchmark selection is twofold. First, an investor must understand that portfolio managers can only execute an investment plan to the extent the market provides them the liquidity to do so. Second, price discovery on the majority of bonds often proves to be difficult; that is, quoted prices and actual trading levels can be fairly far apart.

If the investor chooses to measure manager performance against a benchmark and perform portfolio analyses, it is important for an investor to understand all the idiosyncrasies of a benchmark and the pitfalls of trying to manage against one. In summary, investors must consider specific questions when selecting an index to serve as a benchmark for a portfolio.

What Are Your Investment Goals and Objectives?

Most market participants are investing to meet a specific goal. Individuals might want to save for a house, retirement, or education. Institutions might need to meet specific liabilities. Depending on those goals, an investor might choose a safer or riskier asset class. Even once the asset class has been chosen, many gradations still exist. For example, an insurance company that is looking to increase yield in a fixed-income portfolio might choose to allocate some assets into high-yield bonds. However, insurance companies often need capital to pay off liabilities, and, therefore, capital preservation in their investments is of paramount importance. They might decide that higher-quality, more-defensive high-yield bonds better suit their needs. On the other hand, an investor seeking the highest rate of return might look to the distressed high-yield sector. Careful consideration of specific goals and objectives, along with clear communication of those goals to the manager, will likely point to a benchmark that is the best fit.

The following questions are key issues an investor should consider when choosing a benchmark:

- ***How benchmark-oriented are you?***

  This is an important question that investors might not consider. A manager with a consistent style will likely be able to tell you whether he or she will outperform or underperform in certain market conditions. For example, an aggressive manager who mostly invests in distressed credits will likely underperform in down markets. If the investor is benchmark-oriented and expects the manager to match or beat the benchmark in all market conditions, choosing a broad market benchmark for that manager might not be appropriate. A sub-index or custom index focusing on distressed securities would serve as a more appropriate benchmark for a distressed manager.

- ***How important is liquidity?***

  Although this would likely be decided when initially thinking about investment goals and objectives, liquidity is such a key factor for some asset classes that it must be given further consideration. Even if the liquidity of the investment is not that important—that is, the investor does not need immediate access to his or her money—the liquidity of the asset class can still play a key role in benchmark selection in terms of portfolio construction and price discovery.

- ***Are there other securities you want to allow the manager to own?***

  "Out-of-index" bets include anything the manager holds that is not in the benchmark. Many managed portfolios have a basket of security types other than the target asset class. For example, many high-yield mutual funds hold equities, leveraged loans, government bonds, and/or investment-grade bonds. Usually these are not very large baskets, but they can have a large impact from a portfolio-analysis perspective. Even if the investor is looking for a pure play in one asset class, liquidity again comes into play. The less-liquid the asset class is, the more a manager would likely hold cash, which is also an out-of-index holding that can perform significantly differently than the rest of the portfolio holdings.

Components of a Portfolio Analysis/Performance Measures

A portfolio analysis can comprise many different types of reports or comparisons, and there is no predefined set of required analyses. A large variety of performance measures are available to investors ranging from a simple comparison of portfolio statistics and weightings to a complete performance attribution report. Reports can be on an absolute basis where the portfolio is reviewed independent of any benchmark or they can be relative to an index or even a peer group. Reports can also be for a point in time or for a historical period. Often, it is not the absolute value of a statistic that matters, but rather the relative comparison or even the trend. Some of these types of reports and analyses might include the following:

- ***Total Returns Report.*** Returns are obviously the "headline" number for any investment. It is, of course, the ultimate end goal of any investor—to grow assets and/or fund liabilities. The total return might lead an investor to inquire further about the portfolio especially if the level of return does not meet a hurdle rate or the portfolio greatly under- or out-performs the benchmark. As mentioned before, this is a key reason why appropriate benchmark selection is paramount.

  Reports for actively managed funds can also include returns that are gross and net of management fees. Other types of fees may also be netted out. For fixed-income portfolios, returns may also be broken out into principal return (price change) and interest return (coupon income).
- ***Portfolio Statistics Report.*** One of the most basic, but still very useful and relevant, reports is a snapshot of portfolio-level data points. Besides total return, the relevant statistics differ from asset class to asset class. For corporate bonds, these would typically include yield, current yield, spread, duration, years to maturity, convexity, and average coupon. Some portfolio statistics might not be as relevant to the return, but can still be included, such as number of issues or issuers, total par value, and total market value.
- ***Weights Report.*** This report is a relatively simple breakdown of the portfolio by sector. The number of types of sector will vary by asset class. For corporate bonds, they would typically be

rating, industry, cyclicality, issuer, price bucket, yield bucket, spread bucket, maturity bucket, security type (e.g., floaters, PIKs, zero coupon bonds), security rank (e.g., Senior Secured, Second Lien, Senior), and asset class. For more global portfolios, currency, region, and country of domicile might be added. Ultimately, any statistics that are available at the bond level for all or most of the bonds in the portfolio can be used. This report is useful when wanting to adjust the risk level of a portfolio or understand a trend over time. It is also often done versus a benchmark.

- ***Return Statistics and Relative Return Performance to Peer Group.*** Using a return series, you can calculate a fair number of statistics. Assuming you have a monthly return series for a group of funds, calculating total returns over a specific period is clearly possible (e.g., quarterly, annually). Using the same return series, you can calculate volatility (standard deviation of returns), capture ratios (how much of the upside or downside of the market does the portfolio capture), alpha (a manager's risk-adjusted active return versus a benchmark), beta (a manager's correlation in volatility to the market), drawdowns (a peak-to-trough measure of performance), and others. These calculations can be done versus an index, but they are also often compared with returns of a peer group. Although any portfolio analysis report can be run against a peer group composite, doing so would require obtaining holdings and calculating statistics over time on a large number of portfolios. In lieu of that information, managers and investors can compare return statistics over time for various funds and gain insightful information without underlying holdings.

An example of a key portfolio characteristic that can be calculated using a series of monthly returns is the Sharpe ratio. Developed by Nobel Laureate Professor William F. Sharpe, the Sharpe ratio essentially measures risk-adjusted returns. The higher the Sharpe ratio, the better the manager is at providing return without increasing risk. It is commonly calculated as follows:

$$\text{Sharpe Ratio} = \frac{R^p - R^{rf}}{\sigma^p}$$

Where:

R^p = return of the portfolio

R^{rf} = return of the risk-free rate

σ^p = standard deviation of the portfolio

The formula calculates the portfolio's return over a risk-free rate per unit of volatility. The risk-free rate can sometimes vary, but short-term U.S. Treasury bills are usually used for U.S. assets.[3]

Another statistical technique using a return series for portfolio evaluation is returns-based style analysis (RBSA). Also developed by William Sharpe, RBSA seeks to deconstruct a portfolio in various styles as determined by indexes of various asset classes. It allows the investor to compare the style of portfolio management with a variety of benchmarks to determine which style or asset class the portfolio's return more closely mirrors.

- ***Performance Attribution.*** The culmination of many of the portfolio analysis reports discussed thus far is a performance attribution report. The goal of performance attribution is to explain a manager's active return versus the benchmark (which can be positive or negative) by breaking it down into various market segments, many of which are similar to those described in the earlier Weights Report item. It allows managers and investors to understand what factors aided or hindered performance relative to the benchmark. Therefore, the choice of benchmark is crucially important for performance attribution reports as the information will be distorted if the benchmark is not appropriate. The next chapter discusses the performance attribution in further detail.

Endnotes

1. The company with the smallest market capitalization in the Russell 2000 Index as of July 25, 2013 is Enzon Pharmaceuticals at $81 million with an average daily volume over $250,000.
2. Vanguard High Yield Corporate Fund data is as of July 24, 2013.
3. There are three main articles William Sharpe used to publish his work on the Sharpe Ratio and RBSA. They are the following:
    - Sharpe, William F., "Mutual Fund Performance," *Journal of Business*, January 1966
    - Sharpe, William F., "The Sharpe Ratio," *The Journal of Portfolio Management*, Fall 1994
    - Sharpe, William F., "Determining a Fund's Effective Asset Mix," *Investment Management Review*, December 1988

23

Performance Attribution

Introduction

As evident by the name, the goal of performance attribution is to attribute a portfolio's return to a set of specific factors or sectors. This process aids a portfolio manager or investor to better understand the performance of the portfolio by breaking down the return and identifying sources of return and risk. Attribution can be done on an absolute or relative basis. Absolute attribution, more commonly referred to as return contribution, seeks to identify the largest contributors of return independent of any benchmark. The contribution of Sector A is defined as:

$$\text{Contribution}_{\text{Sector } A} = \sum_{i=1}^{n} W_i \cdot R_i$$

Where:

W_i = weight of asset *i* in Sector *A*
R_i = return of asset *i* in Sector *A*

Although this type of analysis is limited in use, it can still give insight when you are looking for which asset or sector contributed the most to the portfolio. For example, you might want to use a contribution analysis when creating a list of the manager's top ten or worst ten investments.

However, investors and portfolio managers more often want to explain the portfolio's excess return versus a benchmark, which can be positive or negative. This makes sense after the time has been spent on a thorough and extensive benchmark selection process. This analysis, therefore, identifies the sources of return and risk and their effect on the portfolio vis-à-vis the benchmark. For each sector, the decisions a manager makes produce a certain result. These results or "effects" show us whether the sector helped or hindered the return of the portfolio relative to the benchmark. The sum of all the effects equals the portfolio's excess return.

Performance attribution generally follows a sector-based or factor-based methodology, sometimes referred to as a risk, risk-based, or multi-variety approach. The factor-based method attempts to break down and categorize the portfolio's excess return into a set of common risk factors for the asset class plus a residual return. This style of attribution works well for asset classes where specific macro factors can dominate the performance of the portfolio in question as well as for other portfolios composed of the same type of holdings. For example, for many fixed-income portfolios, the risk factors might include interest rates, spreads, and currencies. These types of reports show specific effects that seek to explain how much of the portfolio's return is caused by shifts in the yield curve, twists in the yield curve, and carry. This is then compared with how much of the benchmark's return is based on these same effects to determine which of these factors help or hurt the portfolio relative to the benchmark.

Ultimately, factor-based attribution seeks to explain a portfolio's excess return as a series of decisions on systemic risk factors, which can work well for portfolios driven by the same set of risk factors and with little to no idiosyncratic risk. Because of this, there is general agreement in the equity space on using sector-based rather than factor-based attribution. Corporate bonds and loans, and more specifically high-yield bonds and leveraged loans, are more of a hybrid product—having some of the general attributes of fixed income while exhibiting a fair amount of idiosyncratic risk. Therefore, performance attribution for high-yield bonds is generally calculated using a sector-based attribution methodology.

The standard approach for sector-based performance attribution is commonly referred to as the Brinson method. Brinson and Fachler published their article in 1985,[1] but the basics date back to at least 1972.[2] At its core, sector-based attribution seeks to model a portfolio as a two-step decision process by managers as first determining the weight of a specific sector within a portfolio and then the security selection within each of those sectors. This analysis seeks to decompose the excess return of the portfolio versus the benchmark into three factors or effects: allocation, selection, and interaction. Thus, this approach is sometimes referred to as the three-factor approach to performance attribution.

Allocation Effect

The allocation effect is a measure of the impact of decisions to overweight or underweight particular asset categories relative to a benchmark. According to the Brinson-Fachler model, the allocation effect is a function of both the weighting decision and how that sector performs versus the overall benchmark. Therefore, there are four possible cases, as illustrated in Exhibit 23.1.

Exhibit 23.1 Allocation effect and weighting in a portfolio versus a benchmark

		Portfolio vs. Benchmark Weights	
		Overweight $W_i^P > W_i^B$	Underweight $W_i^P < W_i^B$
Sector vs. Benchmark Returns	Outperform $R_i^B < R^B$	Case 1: POSITIVE Allocation Effect	Case 2: NEGATIVE Allocation Effect
	Underperform $R_i^B > R^B$	Case 3: NEGATIVE Allocation Effect	Case 4: POSITIVE Allocation Effect

Where:

W_i^P = weight of the portfolio in Sector *i*

W_i^B = weight of the benchmark in Sector *i*

R_i^B = return of the benchmark in Sector *i*

R^B = total return of the benchmark

A positive allocation effect results from (a) overweighting sectors that produce greater returns than the benchmark average (Case 1) or (b) underweighting sectors that produce lower returns than the benchmark average (Case 4). Conversely, negative allocation occurs when overweighting sectors with below-benchmark returns or underweighting sectors with above-benchmark returns (Cases 2 and 3). Therefore, the allocation effect for a single period is calculated as follows:

$$\textit{Allocation Effect of Sector } i = (W_i^P - W_i^B) \times (R_i^B - R^B)$$

A follow-up article to the Brinson-Fachler (B-F) methodology was published by Brinson, Hood, and Beebower (BHB) in 1986 and updated in 1991. The new article established a slightly different methodology for allocation. Both models attempt to answer the same question; namely, explaining the excess return by decomposing it into the three effects of allocation, selection, and interaction. However, the BHB study offered a "simplified framework for return accountability" where allocation effect is a function of a manager's decision to weight specific sectors and whether those sectors had a positive or negative return (irrespective of the benchmark return). In other words, the BHB methodology gives credit to a manager who overweights any sector that has a positive return or underweights any sector that has a negative return. Thus:

$$\textit{BHB Allocation Effect} = R_i^B \times (W_i^P - W_i^B)$$

The key difference between the two methodologies is the consideration of opportunity cost. Looking at cases of positive returns in a sector, the BHB allocation effect is positive if the manager is able to overweight any sector with a positive return—in effect making the bogey a zero-percent return. The opportunity cost of investing in a sector is essentially zero. However, the B-F model only returns a positive allocation effect if the manager overweights a sector that itself outperformed the benchmark (e.g., $R_i^B - R^B > 0$). That is, the opportunity cost of overweighting one sector versus another is at least the total return of the benchmark. Therefore, the B-F model is the most commonly accepted among corporate bond managers and investors who use sector-based performance attribution.

Consider the high-yield bond portfolio and benchmark scenario in Exhibit 23.2.

Exhibit 23.2 Weightings and returns for a portfolio and benchmark by rating category

Rating	Portfolio		Benchmark	
	Weight (%)	Return (%)	Weight (%)	Return (%)
BB	15	3.50	45	1.50
B	75	2.00	40	2.00
CCC	10	2.00	15	4.00
Total	**100**	**2.23**	**100**	**2.08**

For this time period, the portfolio was overweight B-rated bonds, nearly double the benchmark weight (75% versus 40%). The total return of B-rated bonds was 2.00% for both the portfolio and the benchmark. Therefore, the BHB model would result in a positive allocation effect of 70 basis points because the manager was heavily overweighting a positive sector:

$$\textit{BHB Allocation Effect} = R_i^B \times (W_i^P - W_i^B)$$
$$= 2.00\% \times (75\% - 40\%) = 0.70\%$$

However, B-rated bonds underperformed the overall benchmark (2.00% versus 2.08%). The Brinson-Fachler method penalizes the manager (in terms of allocation effect) for heavily overweighting a sector that underperformed the overall benchmarking and as a result underweighting other sectors that did better. In this case, the opportunity cost of overweighting B-rated bonds was at the expense of CCC-rated bonds, which return 4% for the period. Thus, the B-F allocation effect is –3 basis points.

$$\textit{Allocation Effect} = (W_i^P - W_i^B) \times (R_i^B - R^B)$$
$$= (75 - 40\%) \times (2\% - 2.08\%) = -0.03\%$$

Although the two methods can result in different allocation effects for a specific sector, the sum of the allocation effects across all sectors will be identical. If the manager is expected to outperform the benchmark, then an effective allocation would be one where the manager overweights the largest outperforming sectors in the benchmark and underweights the largest underperforming sectors. This is how the Brinson-Fachler model views allocation and thus has become the consensus for the calculation of allocation effect among most equity and credit managers and investors. Allocation effect is the only effect that differs between the Brinson-Fachler and Brinson, Hood, and Beebower methods.

Selection Effect

The selection effect measures a manager's ability to choose the best-performing securities within a specific sector and quantify the impact of choosing securities that provide different, hopefully better, returns than the benchmark. The goal is to compare how well the portfolio performed in a sector versus that sector in the benchmark, so the differential in return is calculated. It is then weighted by the benchmark weight; therefore, the portfolio's weight in the sector does not affect the calculation of the selection effect. The weight of that sector in the benchmark does, however, factor into the magnitude of the selection effect—the larger the sector in the index, the larger the effect will be. The selection effect is calculated as follows:

$$\textit{Selection Effect of Sector } i = W_i^B \times (R_i^P - R_i^B)$$

Using the hypothetical high-yield portfolio in Exhibit 23.2, we can calculate the selection effect for BB-rated bonds as follows:

$$\textit{Selection Effect of BB-rated Bonds}$$
$$= 45\% \times (3.50\% - 1.50\%) = 0.90\%$$

The manager was able to choose BB-rated bonds that had returns of 3.5% during the period versus only 1.5% for BBs in the benchmark. This resulted in a selection effect of 90 basis points. Although the weight of BBs in the portfolio does not factor into the equation, it would stand to reason that the smaller the weight of that sector in the portfolio, the easier it would be for a manager to buy only the "best of the best" BB-rated bonds and achieve a far superior return.

The selection effect will be zero for all sectors to which the benchmark has no exposure; that is, out-of-index bets. For example, if an investment-grade portfolio manager decides to purchase some high-yield bonds, then the portfolio will have exposure to a sector the benchmark does not have—assuming, of course, the benchmark is an investment-grade bond index. This can occur rather frequently depending on what sectors are used for the attribution report, but because most corporate bond portfolios hold at least some cash and the benchmarks generally do not, the cash would be considered an out-of-index bet and would, therefore, have a selection effect of zero.

Additionally, when the index weight of a sector is not zero, but the portfolio does not hold any of those securities, the portfolio return in that sector would simply be zero. Therefore, because the portfolio weight in the sector (W_i^P) is not explicitly an input in the selection effect formula, it is theoretically possible to calculate a selection effect using a portfolio return of zero ($R_i^P = 0$). However, logically this does not make sense. That is, how do you assess the manager's ability to select credits in a sector when there are none in the portfolio? Therefore, many attribution systems zero out the selection effect if the portfolio does not have any securities in a given sector.

Interaction Effect

The interaction effect is probably the most conceptually difficult to comprehend. It is meant to measure the combined impact of allocation and selection in the portfolio. Attribution and selection effects are fairly intuitive—how well the manager chose sectors and how well the manager chose specific credits. Both of those are decisions of the manager, whether intentional or not. However, the interaction effect does not measure an explicit decision of the manager, but rather, as the name suggests, how the two decisions of allocation and selection interacted. This might be somewhat of a catchall term. In fact, in the original BHB article, it was referred to as simply "other."

The formula for interaction effect combines weight and return differences. Thus, interaction effect is calculated as follows:

$$\textit{Interaction Effect of Sector i} = (W_i^P - W_1^B) \times (R_1^P - R_1^B)$$

Comparing this formula to the formulas for the other two effects, you will notice this formula takes into account allocation (difference in weights) as well as selection (difference in returns). Therefore, the interaction effect will be positive if the manager chose to overweight a sector and managed to select outperforming credits. In other words, the manager is rewarded with a positive interaction effect if he or she has credit expertise in a sector *and* the conviction to overweight it. The interaction effect is also positive if the manager chose to underweight a sector and selected worse credits. For example, the manager is unfamiliar with a particular sector and thus would have difficulty choosing the best credits, but he or she has the foresight to buy less of that sector. Any other combination would lead to a negative interaction effect.[3]

Interpreting the Total Effect

The goal of performance attribution is to seek to explain the portfolio's return versus the benchmark. Consider the performance attribution report in Exhibit 23.3. The total effect for cyclical bonds in the portfolio was a −10 basis point detraction from the overall portfolio, which is obtained by summing the allocation, selection, and interaction effects. Reviewing the cyclical line, the portfolio was overweighting the sector, while underperforming. Therefore, the allocation effect is positive (overweighting an outperforming sector in the index), the selection effect is negative (portfolio return underperformed benchmark return in that sector), and the combination would yield a negative interaction effect (overweighting an underperforming sector).

For the cyclical sector:

$$\textit{Allocation Effect} = (W_i^P - W_i^B) \times (R_i^B - R^B)$$
$$= (60\% - 50\%) \times (1\% - .54\%) = 0.05\%$$

$$\textit{Selection Effect} = W_i^B \times (R_i^P - R_i^B) = 50\% \times (0.75\% - 1\%) = -0.13\%$$

$$\textit{Interaction Effect} = (W_i^P - W_i^B) \times (R_i^P - R_i^B)$$
$$= (60\% - 50\%) \times (0.75\% - 1\%) = -0.03\%$$

Exhibit 23.3 Illustration of allocation, selection, and interaction effects by sectors

	Portfolio		Benchmark	
	Weight (%)	**Return (%)**	**Weight (%)**	**Return (%)**
Cyclical	60	0.75	50	1.00
Defensive	35	1.25	40	0.25
Energy	5	0.20	10	-0.50
Total	100	0.90	100	0.54

	Allocation Effect	**Selection Effect**	**Interaction Effect**	**Total Effect**
Cyclical	0.05	–0.13	–0.03	–0.10
Defensive	0.01	0.40	–0.05	0.37
Energy	0.06	0.07	–0.04	0.09
Total	0.12	0.35	–0.11	0.36

Note: Totals may not appear to sum correctly due to rounding.

The overall portfolio outperformed the benchmark during the period by 36 basis points (0.90%–0.54%), which is shown as a sum of the total effect for each sector on the lower right. The sum of the total effects for Cyclical, Defensive, and Energy bonds equals the 36 basis point outperformance on a relative basis to the benchmark.

Two-Factor Approach to Performance Attribution

Because the interaction effect is not a direct result of an active decision by the manager, there is a fair amount of disagreement in the attribution community as to whether interaction effect should be included at all. This approach is preferred by some analysts, and some reports group selection and interaction by default thus making it the *de facto* approach for those reports. The two-factor performance attribution is fairly similar to the three-factor approach, but leaves out the interaction effect. This is essentially done by combining selection and interaction effects. Mathematically, this can be done by changing the weighting factor in the selection effect formula. In other words, you would use the portfolio sector weight in lieu of the benchmark sector weight in the formula, as shown next. The allocation effect (using either BHB or B-F models) would not change.

$$Selection\ Effect_{Two\text{-}Factor} = W_i^P \times (R_i^P - R_i^B)$$

In both the two-factor and three-factor approach, the crux of the analysis remains the difference in returns ($R_i^P - R_i^B$) as the representation of the manager's ability to pick credits. The arguments for and against the two-factor model are numerous and lengthy. A rudimentary oversimplification of the argument against using the two-factor approach is that the selection effect should weight the effect by the benchmark return to isolate to credit selection skill. An equally oversimplified argument for two-factor analysts argues that the selection effect should measure the effect of credit selection on the portfolio and thus the weighting factor should be the portfolio weight. The interaction effect may still be an effective tool in the overall analysis, but would be considered a subeffect of the total selection effect. Clearly, both approaches have their merit and the discussion goes beyond the scope of this book.

Challenges of Sector-Based Performance Attribution

Like nearly all analyses, sector-based performance attribution has its advantages and disadvantages. Because the goal is to calculate each of the effects by separating the portfolio into different sectors, one of the greatest benefits of sector-based performance attribution is nearly infinite flexibility in sector choices. However, in order to perform the analysis, all risk factors must be divided into specific groups even if it may be difficult to separate the risk into distinct groupings.

Factor-based attribution, on the other hand, can analyze the effect of a specific risk factor as a whole. For example, even for high-yield bonds, you might want to analyze the effect of interest rates. Factor-based attribution could calculate a "curve effect" while sector-based attribution would look at buckets by duration, maturity, coupon, or another characteristic that represents rate risk.

Another criticism of sector-based attribution is that it can confuse, or at least blend, conscious allocation decisions with unintentional ones. For example, a distressed credit manager may look at credit based solely on fundamentals. As such, the portfolio may have a significant overweight in higher coupon bonds. In a rising rate environment, a sector-based attribution report using coupon groupings may show a positive allocation effect to the lower coupon groupings (underweighting an underperforming sector), but this positioning was not a conscious decision of the manager.

Lastly, a key criticism of performance attribution in general is that it is entirely based on ex-post results and, therefore, can be limited in its usefulness. In other words, a performance attribution report tells us what helped or hurt the portfolio relative to the benchmark *in the past.* Of course, the market environment in the next period might or might not be similar to the last period. However, understanding how the portfolio is positioned and the relative returns for each sector can help managers bridge their portfolio construction with their views on market conditions and the overall economic environment with the hope of shifting their portfolio to match their views.

Endnotes

1. There have been three main articles for the basis of sector-based performance attribution published by Gary P. Brinson et al in 1985, 1986, and 1991. They are the following:

    - Brinson, Gary P., and Nimrod Fachler, "Measuring Non-U.S. Equity Portfolio Performance," *Journal of Portfolio Management*, Spring 1985

- Brinson, Gary P., L. Randolph Hood, and Gilbert L. Beebower, "Determinants of Portfolio Performance," *Financial Analysts Journal*, July/August 1986
- Brinson, Gary P., Brian D. Singer, and Gilbert L. Beebower, "Determinants of Portfolio Performance II: An Update," *Financial Analysts Journal*, May/June 1991

2. In 1972, The Society of Investment Analysts published a report titled "The Measurement of Portfolio Performance for Pension Funds," which discussed some of the ideas of analyzing portfolio performance as two components—selection of stocks and selection of sectors.

3. For further information about Interaction Effect, see Spaulding, David, "Performance Attribution: A Powerful Tool for Identifying Sources of Investment Performance," Advent Software White Paper, 2013.

Closing Comments on Section IX

Portfolio analysis and portfolio attribution are major areas of data analytics in all of investing. As outlined in this section, there are numerous nuances to applying it to corporate debt. Most notably is the many ways that you can analyze the attribution, using both aspects that are typically used in the equity markets as well as attributes that are typical of fixed income. Part of the analysis is to determine what drove performance. When trying to utilize this analysis to be more forward looking, you must have a sense of what type of cycle you are expecting and then you can examine portfolio attribution from the past during similar cycles.

Portfolio attribution is certainly studied by portfolio managers and asset allocators, but trading desks at sell-side firms could also benefit greatly in their risk management and performance attribution by applying some of these techniques to their trading operations. For example, they could add buckets to look at performance by their own underwritten transactions versus those by other firms or by the average age of positions, both critical points for them to understand.

Section X
The Future of Data Analytics and Closing Comments

By Robert S. Kricheff

I believe that data analytics in the corporate debt markets is still fairly early in its development. Improved use of computer systems and improved market data are both likely to lead to the increased use of more advanced techniques in the corporate debt markets.

There is no question in my mind that the complexity of the data needed to monitor corporate debt and the lack of consistently good pricing data have both slowed the evolution of analytics. However, even with these issues, there should be wider use of analytics in the debt markets. Some newer developments outlined in the next chapter might help lead to broader and different uses of analytics for corporate debt.

24

Some Thoughts on the Future of Data Analytics in Corporate Debt Markets

Although the growth in the size of the corporate debt markets in recent years has been revolutionary, the development of analytics for the market has been evolutionary at best. I expect the demand for data analytics will increase in coming years, but I believe more of the investment will have to come from money managers and third-party vendors as opposed to the investment bank/broker-dealer side, where much of the investment has come from so far.

The techniques and methods currently used are still relatively simple, but I think it is quite likely that they will become increasingly sophisticated and complex. This chapter outlines some of the areas where analytics may expand and some of the items that are holding back some of these developments. Many of the developments would not necessarily be revolutionary in how or what people do, but would accelerate the response time and speed at which investors, traders, and even asset allocators could react.

Bond Data and Fundamental Data

One of the key evolutions that needs to take place is a better linking of bond data analytics and credit metric analytics that are derived from analysts. Increased automation in this area should help speed up analysis and decision making.

Currently, many systems that include robust information about bond data and bond metrics—such as yields, pricing, and duration—cannot connect or link with systems that show key credit metrics on the same bond issues—such as leverage ratios or revenue and earnings trends. Upon entering the next credit crunch cycle, there will be demand for models that run more alerts and highlight potential covenant violations, alerting investors immediately when financial results are updated, when headroom on maintenance tests comes close to being violated, or when prices decline meaningfully with minimal changes in fundamental data. This would all be more easily accomplished with better links between these types of data.

Third-Party Vendors of Financial Data

There may be a push to use financial data and calculations from third-party sources. Many of these services are quite good and are widely used. This approach might initially save money and appear to save time, but it has many risks. There is a real danger in relying on these too much or making investment decisions based on this information.

These types of services can be very useful to get a quick snapshot of certain aspects of a company, and can sometimes be useful to do some quick sorts and queries. They can also be helpful to get a quick picture of a company that might not currently be on an analyst's coverage list or in the universe of names a trader or portfolio manager currently looks at.

However, as good as these databases of financial data get, they are highly likely to do some vital parts of the analysis differently than you or your analyst would. As long as Murphy's Law will be quoted, the time that something is missed using a third-party vendor of financial data, will be the time that it will cost you the most money on a trade or investment decision.

The problem does not lie in the possibility of a mistake being made, because you or your analyst is capable of that, too. The problem lies in understanding the nuances and structures of how you want to look at credit data and the details (and warning signs) that are derived from going through financial statements yourself. For example, different people include different items in an adjusted EBITDA calculation; some add back a private equity company's fees and some do not, some give credit for one-time restructuring costs and some do not, and some add back the charge but net out the cash expenses. These differences can be material. You can go through many other key items as well, which might not be caught by a third-party vendor, such as one-time charges, items on the flow of funds statement, or even how total debt is calculated for many companies.

Prioritization of processing timely releases of financials might vary greatly, too. You also need to go through the financial statements to understand a company. Although many of the nuances that you uncover in crunching the numbers yourself might not show up in a large data study of shifts in leverage ratios, what you miss by not having your own people go through the financials could be crucial. For example, if you just take the changes in working capital from a database and put it in your spreadsheet, it might look like only a modest change. You might miss the nuance that inventories have risen sharply during the last two periods while payables have, too, which could be a telling fact about the direction of the business. Additionally, after you have looked at a company for a while, there is a sense of its financial statements and unusual things (both good and bad) catch your eye; even a change in how the company reports business segments can prove to be significant.

Too much reliance on third-party financial spreadsheets can get the best-intentioned analysts away from reading the actual filings. In reading through those statements, and especially the footnotes, you often find informational treasure troves about the credit quality of a company.

Growing Use of Word Recognition

Improved word search techniques and word recognition techniques could also be increasingly used in reviewing quarterly financial statements and for other Securities and Exchange Commission (SEC) documents.

For example, perhaps you are worried about a company selling a division of its business. It would be extremely helpful if there was a system that could alert you to any news or phrase about this in any company press release or public filings as soon as it came out, especially as sometimes this information can be buried in very long financial filings.

Word recognition techniques could also be used in helping to map and organize covenant analysis. Key words and defined terms are often spread throughout the documents, but are needed when reading a clause that is placed far away in the document from the meaning of the defined terms. Linking or being able to pull those definitions into the clause you are reading could be helpful and time saving.

Covenant Analysis

Covenant analysis in general is very tedious and even minor automations or improvements could prove to be very helpful.

Covenants are an important differentiation among debt instruments. This is particularly true in the high-yield bond and the leveraged loan market. Covenants often come into play in a specific event scenario and this is difficult to anticipate. For example, if a company is making an acquisition, is it limited by its leverage test in a debt incurrence covenant, or are there carve-outs that would allow additional borrowing under the debt investment covenant, or does the definition of EBITDA in the leverage test allow for extensive pro forma adjustments in the case of an acquisition (and is that definition in a completely separate section of the prospectus from the debt incurrence language)? For reasons such as the example in the prior paragraphs, I have always found it useful to actually diagram covenants when they are critical to an investment, including key ratios and calculations in each box of a decision tree-like diagram. Systems could evolve where a covenant diagram could also have automatic calculations as you update the financial information.

There have been attempts to quantify and score covenants, but thus far, they have not been rolled out widely and I believe will have limited acceptance because of the extreme subjectivity in these systems and the limited value of them. The problem is that covenants can be extremely nuanced, and I believe too nuanced for any scoring system to be helpful. Scoring also does not help in analyzing specific events, as described previously.

Multiple Scenario Analysis

As data analytics becomes increasingly embedded in the mindsets of more corporate debt market participants, the use of multiple scenario analysis will likely increase. I think everything from simulated model portfolios to individual scenarios for specific credits and their underlying debt instruments will be more commonly run.

These types of scenarios could be run for pro forma mergers or perhaps to run total return analysis on refinancing scenarios for an entire capital structure. There are already many places that have programs and spreadsheets that can run these periodically and on a one-off, ad hoc basis. However, over time I could envision these reports coming out every morning to portfolio managers or traders about all of their positions or selected positions that they want to monitor as well as to capital markets and investment bankers about their clients' capitalizations. This could provide valuable insights into potential positive event risk and upside/downside scenarios for the related debt securities.

Data Mining

The use of data mining will likely increase. Much of this is done already by fixed-income, macro style hedge funds to track relationships between asset classes as well as between some of the structured products and their underlying assets.

Trades are put on, such as buying an ETF and shorting a basket of the likely bonds that are being purchased in the ETF to find where arbitrage opportunities are. I think data mining will be more widely used to analyze opportunities within a given market between various tiers and subsectors of the market. As this evolves, there will be market participants who look to develop relationship trades between some of these subsectors based on this data.

Use of Big Data and more data mining techniques will certainly evolve in the corporate debt markets, but this will take time. Improvements in linking bond trading data and financial metric data into a single platform should help to increase the amount of ways that data mining techniques can be used, but better pricing data will also be needed to help the use of these techniques.

Indexes

Indexes have been one of the driving forces of data analysis thus far in the corporate debt markets. Although we have seen evolution in the use of these indexes, in structured products, increases in the actual capabilities of the indexes have been somewhat limited.

Some of the possible near-term developments in index products include increases in customized sorts and queries, increased real-time data, an increase in the amount of global issuer data, and the ability to automatically run hedges on both fixed to floating rate and among currencies, as both leveraged loans and global markets have become increasingly important asset classes.

More detailed and broader capabilities in leveraged loan indexes would be widely welcomed. However, the difficulties in gathering data in the leveraged loan market are likely to persist. Trading is very sporadic, often the agent bank controls the trading in a loan (often permission is needed to transfer the claim). Building indexes and databases can be trickier without easy identifiers (e.g., such as the CUSIP system in bonds[1]). Additionally, terms of the note can change with a limited amount of information distributed; that is, only current holders of the debt are informed about changes that have been approved by a majority of the holders.

As discussed earlier, all of the widely used indexes have been developed and are managed by investment banks. With all the current restrictions and increased regulatory pressures on these entities, it is likely that more IT resources are being spent on compliance than on client service or improving index products. Additionally, no dominant index in these markets can be viewed as the go-to index, the way there are some in the equity markets. Therefore, there is much debate as to which one to use. Development of a dominant index for the leveraged debt asset classes could prove to be valuable, but I believe this would most likely evolve in an index developed by a third-party vendor.

Pricing and Liquidity

One thing that definitely gives people pause before they invest more in data analytic systems is the quality of the information on prices in corporate debt markets. Any improvements on trading transparency and volumes would increase the value of data analytics.

Coupled with the difficulty of getting good prices is the difficulty in execution. Most bonds and loans have very limited liquidity. If you invest heavily in analytics and the systems show you that you should trade out of a given bond, and then you cannot execute that trade in large size, you have to question the value of the investment in the analytics.

Automation and this shortfall in liquidity will likely force an increase in electronic trading platforms. These will likely never end up resembling the activity and transparency of the equity markets, but could greatly improve the corporate debt markets. There is increasingly limited capital among broker-dealers to buy in and hold debt, so they must find a buyer on the other side and this limits how much can get done at a given price. This current environment certainly creates opportunities for the development of electronic platforms that may offer more peer-to-peer trading.

A Final Concern about the Future of Analytics

Thus far, much of the development of data analytics and the tools used in data analytics have been developed on the sell-side of the business. With all of the regulatory changes for the investment banks in most countries throughout the world, significant resources are being spent on meeting regulatory reporting requirements. I am worried that these developments are leading to much less investment in data analytics in the corporate debt markets, when they could be increasingly used to help manage risk. I believe the willingness to invest in these analytics by other participants in the market is varied at best. This leads to the concern that there may be a slowdown in the developments in these areas. Foresighted asset management firms will clearly invest in this area to develop advantages in the markets for their customers.

Endnote

1. The CUSIP system assigns a unique number to each debt security that is issued.

25

Closing Remarks

Data analytics is an invaluable tool to operating in the corporate debt markets. Although everyone in the market directly or indirectly uses some of these tools, there is a huge advantage to those who invest in and incorporate these tools in a disciplined fashion. Proper use of data analytics can help market participants to more appropriately allocate their resources and to react quickly and correctly when trends start to shift.

Similar to most analysis, data analytics is not a final answer to all questions. There are many difficulties in utilizing data analytics in the corporate debt market; there are many traps that you can stumble into; and there are many shortcomings in the conclusions. The corporate debt markets require significantly more information in their related databases to be able to analyze the market than the equity or government bond market does. Additionally, pricing data and trading liquidity to be able to execute strategies are both severely limited and spotty throughout the markets. Poor pricing and a constantly changing universe of constituents within the credit universe makes historical analysis as well as structural analysis fraught with potential traps and mistakes. Finally, data analytics, like credit analysis, can be used to make projections. As precisely as these projections may be modeled and rationalized, they are still based on historical data and cannot help but have some aspects of subjectivity in the results. Remember: When analyzing and reviewing historical data on any financial markets, *if you begin to accept patterns as rules, an exception will eventually crush you.*

This book outlined many of the tools currently used in the marketplace. It also outlined some of the tools that might be used in the future. Do not believe that these are the only ways to use some of these tools for data analytics and do not believe that there are no other tools that can be developed or adapted to be used in this market. Many of these analytics are fairly simple and I believe significant steps forward could be made; in particular, more theoretical work on reward per unit of credit risk would be particularly interesting and might possibly evolve into more interesting analytics for the market.

Picking the right credits and debt instruments to meet your needs is still ultimately the goal of most market participants. However, this work is best not done in a vacuum. The data analytics outlined in this book hope to fill that vacuum and make the entire process more efficient, focused, and disciplined. I hope you will think about and develop many more ways to analyze these markets.

Index

A

B

C

E

F

G

H

I

J-K-L

M

Q-R

S

T

U-V

W

X-Y-Z